Assessment and Communication of Risk

Eric Liberda · Timothy Sly

Assessment and Communication of Risk

A Pocket Text for Health and Safety Professionals

Springer

Eric Liberda
School of Occupational & Public
Health and School of Graduate
Studies
Toronto Metropolitan University
Toronto, ON, Canada

Timothy Sly
School of Occupational & Public
Health and School of Graduate
Studies
Toronto Metropolitan University
Toronto, ON, Canada

ISBN 978-3-031-28904-0   ISBN 978-3-031-28905-7   (eBook)
https://doi.org/10.1007/978-3-031-28905-7

This Springer imprint is published by the registered company Springer Nature
Switzerland AG
The registered company address is: Gewerbestrasse 11, 6330 Cham, Switzerland

To you, dear reader. May the volume you are holding in your hands provide a guiding light as you assess the risks and exponents of outrageous probability.

Preface

Although this pocket reference was developed primarily as a resource for those tasked with assessing risks and communicating those risks in the fields of public health and occupational safety, other technical and professional personnel in any related health or safety field will find it useful. The focus is mainly on the basics of quantitative risk assessment, probabilistic risk assessment, and the communication of risks to the media and general public, but a brief introduction to several methods of qualitative risk assessment is also provided, with suggestions and links to other resources for those techniques.

Risk-based decision-making is a frequently-heard term in all manner of organizations and agencies in both public and private sectors. The assessment of the hazard, the probability of a hazardous or dangerous event taking place, together with the estimation of the magnitude and consequences of that event, even though it has not yet happened, is the basis for countless decisions in applied health and safety, resource allocation, accountability, loss-prevention, decision-making under uncertainty, needs assessment, and public acceptance.

The meaning of "risk," of course, can mean something different to each profession or discipline, and indeed to each individual, based on their background and previous experience. This guide addresses human health and safety risks in domestic, occupational, and environmental settings, as well as in emergencies. We discuss the general types and categories of risk, demonstrate the methods used to calculate those risks in a range of settings, and illustrate the interpretation of outcomes, such that risks can be reduced, managed, or controlled more effectively.

The material presented here avoids many of the theoretical underpinnings and instead provides a step-by-step approach to completing and implementing a basic risk assessment for personnel in health and safety fields.[1] The calculations and worked examples included in this volume are straightforward and should present no difficulty if a simple calculator is available. Familiarity with probabilities and expressing exponential functions is quickly re-learned or refreshed. Also provided is a section on the correct use and interpretation of the relative risk and its substitute, the odds ratio, essential tools in all areas of research and investigation.

While other texts can be found presenting various levels of "how-to" information on risk assessment techniques, few also include the difficult issue of how the public and media perceive those risks, and how the results of those assessments are best communicated, explained, and discussed with the people who are expected to make decisions about their health and safety. With this in mind, the final chapter in this work (Chap. 6) offers an introduction to understanding risk perception and provides a wealth of useful tips and advice in preparing for communication activities, face-to-face, in town-hall settings, and via media channels. A number of case studies in communication have been summarized, and for those seeking more in-depth work on this essential subject, we highly recommend Regina Lundgren & Andrea McMakin's sixth edition of *Risk Communication* (2018), a work that is without equal in its practical approach and comprehensive scope toward assisting health and safety professionals in communicating risks.

The book is arranged into six chapters:

1. ***Terminology, calculating probabilities, definitions, and the general methods*** used in assessing risk, together with an introduction to the ways risks, are classified for various purposes. A

---

[1]For a more detailed mathematical understanding of the assessment process, authoritative reference texts are available (e.g., Paustenbach 2009).

revision of the purpose, application, and interpretation of the relative risk, and its substitute, the odds ratio, is included, together with several examples.

2. *Catastrophic risk assessment* (also called *probabilistic risk assessment* or PRA) is introduced as an objective method of predicting and measuring the likelihood of accidents, disasters, and other significant events that may or may not take place. Analytical methods include the formulae, contingency tables, Venn diagrams, and especially probabilistic trees. Numerous worked examples are shown together with solutions and case studies.

3. *Chronic risk assessment* (also called *human health risk assessment* (*HHRA*) or *quantitative risk assessment* (**QRA**)) is the process of predicting and measuring the risk of carcinogenic and non-carcinogenic effects due to chronic exposures to substances in the air, food, water, workplace, or pharmaceuticals, over an individual's lifetime. These methods draw from statistics, engineering, physics, chemistry, biology, medicine, toxicology, and epidemiology. Numerous worked examples are shown together with solutions and case studies.

4. *Qualitative methods of risk analysis* are described, and compared, not as a comprehensive resource, but as an introduction to six of the more commonly used methods, providing a useful overview of their potential applications, along with procedural diagrams and two case studies.

5. *Risk assessment in food safety and foodborne illness* has become more urgent as the scale of food production and distribution takes on global characteristics. Even small errors or oversights can result in innumerable cases of illness and even deaths. Some risk assessment techniques from industry and institutions have been successfully applied to food, with the scope of concern starting with raw ingredient production all the way to processing, transportation, distribution, and consumption.

6. *Risk perception* and *risk communication* assumes that the predictions, estimates, and probabilities have all been calculated, leaving us with the task of informing the stakeholders and helping people to make decisions about reduction or

elimination of those risks. This involves not only the *comparison, and explanation* of risks, *in context,* to colleagues, management, government, media, and especially with the public, but also understanding how people perceive risks and threats to their health and safety. Poor risk communication can, and often still does, inflame the public, and contribute to public distrust toward industry, government, and authority in general.

Finally, beyond the scope of this book are the many aspects of *risk management* which involve economics, resources, policy issues, planning, theory, needs-assessment, administration, program evaluation, and the establishment, interpretation, and enforcement of legislation and standards. A brief discussion on risk management appears as Sect. 1.2, and readers are encouraged to refer to the international standard (ISO 31000) on Risk Management.

Toronto, ON, Canada   Eric Liberda
                                                         Timothy Sly

Reference

Paustenbach, DJ. 2009. Human and Ecological Risk Assessment: Theory and Practice (Wiley Classics Library). Wiley Publishing. ISBN: 978-0-470-25319-9.

Acknowledgment

We are indebted to our families and colleagues for encouragement and logistical support during the creation of this somewhat risky work.

Contents

List of Figures

About the Authors

Eric Liberda, PhD, is a professor in the School of Occupational and Public Health at Toronto Metropolitan University. He holds qualifications in toxicology and risk assessment from New York University (USA), RMIT (Australia), and the University of Waterloo (Canada). Dr. Liberda serves on government advising groups, conducts research related to human exposures, and has dedicated his life to helping others.

Timothy Sly, PhD, is Professor Emeritus in the School of Occupational and Public Health at Toronto Metropolitan University. He holds public health qualifications from UK and Canada, and graduate degrees in Epidemiology (University of Western Ontario) and Risk Studies (Teesside University, UK). He has also taught and worked in Brazil, Philippines, Taiwan, and the Caribbean Region, and served on expert committees for food and waterborne diseases.

Risk: Scope, Definitions, and Terminology

Abstract

We begin with an introduction to the meaning, language, terminology, scope, definitions, and categories of risk in the health and safety settings. Many of the measurements and outcomes of risk analysis are in the form of probabilities, and for this purpose, the calculation of probabilities is reviewed, not in academic detail but sufficient to enable health and safety professionals to use available data appropriately in generating probabilities of events happening, or probabilities of fatalities and other unfortunate outcomes. While initial illustration of these methods often employs cards, coins, and dice, the reader will find they are immersed in practical applications by the end of the chapter. A comprehensive revision of the relative risk and its substitute, the odds ratio, have been included because of the ubiquitous nature and application of these important measures in all areas of research and evaluation.

The original version of the chapter has been revised. A correction to this chapter can be found at https://doi.org/10.1007/978-3-031-28905-7_7

1.1   The Scope of Risk Assessment

A summary and analysis following an accident, leak, spill, release, fatality, or even a "near-miss" incident identifies and evaluates the contributing factors and the role played by each, either directly or indirectly. This is the case history, and while it certainly provides valuable data and insights to inform future risk assessments, the case history examines what has already occurred. The risk *assessment* ventures to predict the probability and magnitude of what has not yet happened.

You may have encountered probability concepts working with statistics and may be familiar with statements following an analysis such as "P > 0.05 or P < 0.001," or terms involving chance or odds. In this chapter, we will become more familiar with the use of probabilities in measuring risk because of course we are estimating outcomes or events that have not yet happened and estimating the likelihood or probability that they *might* occur, together with predictions of the consequences if they *do* occur.

As to the importance and position of risk assessment in human health and safety, we can only note the increasing pressure in recent years to predict, measure, and prevent or attenuate accidents, spills, environmental damage, injuries, fatalities, and other unwanted events. Decreasing availability of resources, increased competition for funds, public demands for due diligence in safety, and accountability for expenditures and future planning have all contributed to raising the profile of risk assessment. In the fields of health care, public health, and occupational health and safety, the prefix "*risk-based-*" *is* increasingly applied to the decision-making process in operational departments, agencies, and organizations. This guide is designed to provide additional familiarity with a wide range of assessment methods and to demonstrate a range of toolbox techniques. But first, a word about the relationship between risk assessment and risk management.

1.2   Risk Management

While risk assessment involves data gathering and the analysis of the scientific evidence about risks, risk management typically involves regulatory decisions, as well as the activities involved in planning and implementation of the remediation or reduction of the risk, the finances, resources, policies, and decisions about the extent, and priorities, of appropriate risk reduction measures.

A "conceptual distinction" is generally maintained between risk assessment and risk management, on the grounds that risk assessments, subsequently used for risk management decisions and options, should not be inappropriately influenced by resource issues, or political or policy expediencies or the preferences of those making the management decisions.

> ...*It is imperative that risk assessments used to evaluate risk-management options not be inappropriately influenced by the preferences of [decision makers]*
> **Science and Decisions: Advancing Risk Assessment** (NRC 2009)

Valuable resources for risk management are available through the International Standards Organization in the form of the *ISO 31000 Risk Management Guidelines* (2019). This provides set of internationally adopted principles for managing risk and frameworks for achieving targets based upon best practices. It can be adopted and implemented by organizations of any size or industrial sector, with the goal of increasing the likelihood of achieving objectives, improving the probability of identifying opportunities, obstacles, and threats, and using resources to control risk more effectively.

ISO 31000 has an independent status, and hence cannot be used for HACCP-certification purposes. It can, however, be used to guide risk-audit programs, both internal and external. ISO 31000-certified organizations can contrast their risk management practices with international norms.

Notwithstanding the *arms' length* relationship between assessment and management of risk, the risk characterization step in Chap. 3 incorporates and compares options for reduction of risks that arise from chronic intakes through air, water, or food. Also, the communication of risk has components essential to both risk management and risk assessment and, for this reason, is included as Chap. 6 of this book. Figure 6.1 depicts graphically the role of risk communication vis-à-vis risk assessment and risk management.

1.3   Basic Definition of Risk

The language of risk varies across applications and disciplines. "*Risk*" can be expected to convey different meanings and applications to a physician, an engineer, an epidemiologist, an insurance company, a stockbroker, or an occupational hygiene specialist. And the meaning of "*risk*" varies in even more important ways between "experts" and laypersons. In a popular sense, "risk" is just the possibility of an adverse outcome, or of suffering harm or loss, and an element of "risk" can be found in any endeavor or event; risk is ubiquitous. But if we want to measure risks, and compare risks, the technical definition of risk remains:

$$\textbf{Risk} = \textbf{Probability}\,(\textbf{of event happening})$$
$$\times\, \textbf{Magnitude}\,(\textbf{should it happen}): RISK = P \times M$$

The harm, injury, or loss must be defined of course, as must the time interval in which the harm, injury, or loss could take place (e.g., fatalities per year, injuries per month, or person-years lost due to accidents).

Figure 1.1 shows this type of calculation comparing hypothetical injuries that vary in severity (e.g., number of days lost from work), as well as in the likelihood (probability) of these injuries happening: A back injury results in a mean of 15 days lost, but the risk of it happening to a specific worker in any year is only 10%. Shoulder injuries will typically result in the loss of only 5 days work, but the likelihood of the injury occurring per worker is once in 5 years (one-fifth or 20% chance of it happening annually to a worker). Hand injury is more common (one chance in two for any person in a year) and results in an average of 3 days lost each time.

The severity of the injury (its "magnitude," here defined as days lost from the workplace) is multiplied by the annual proba-

Injury type	Severity of outcome (average person-days lost per injury)	Annual probability of injury per person	Risk (person-days lost per year)
Back	15	0.10	1.5
Shoulder	5	0.20	1.0
Hand	3	0.50	1.5
Back and hand injuries have similar annual risk (1.5 pd/y) but shoulder injury is less (1.0 pd/y)			

Fig. 1.1 Comparing injury risks using R = P × M

bility of that type of injury. It is worth noting that the units, when used correctly, should all "balance." This is highly recommended for all such calculations. For example:

$$\frac{15 \text{ days}}{\text{injury}} \quad \times \quad \frac{0.10 \text{ injury}}{\text{year}} \quad = \quad \frac{1.5 \text{ days}}{\text{year}}$$

1.4 Describing Risks Using the Language of Probability

Most people want a simple answer to their question: *"Will this food additive give me cancer?"* or *"Is this bridge safe?"* We cannot answer such questions in a *"yes/no"* format, because there is always the possibility that even the safest bridge may fail or, conversely, that an exposure to a known carcinogen may never cause an injury or cancer. This is one of the distinctions between cancer-causing exposures and exposures to noncarcinogenic toxicants. If an adult consumed 400 mg of sodium cyanide, we would expect this to be a lethal dose in every instance. But, we all know the rare one or two people who have smoked 25 cigarettes a day for a lifetime and yet somehow have managed to avoid lung cancer. Figure 1.2 offers another perspective on $R = P \times M$.

By measuring the relevant factors, we can estimate *how likely* the cancer death will be, or *how likely* it is that the bridge will fail.

Analysis challenge: .... Two sources of electrical power for the community are being compared.

- **A liquified natural gas (LNG) plant** near a mobile home park has an estimated risk of the main storage vessel exploding once in 40 years, with an estimated 20 deaths expected as a result.

- **A coal-fired generating plant** is the direct "cause" of 2 deaths per year due to emphysema, bronchitis, and cancer from air pollution. Which one has a greater annual risk, and by how much?

Comparing *annual* risks of deaths, the LNG facility causes 20 deaths "spread" over 40 years, = 20d / 40y, or 0.5 deaths per year. Meanwhile, the coal plant is responsible for 2 deaths per year, so it is actually *four times* as risky compared to the LNG facility!

Fig. 1.2 Comparing fatality risks

For that, we need to become familiar with measuring and expressing probabilities: *"What is the likelihood or probability of that event happening?"*

Note that while *probability* is always expressed as a value between zero and one, *risk* is often expressed in "real-world" measurements (e.g., predicted "events," "injuries," "accidents," "deaths," "days lost," etc.).

1.5 Categorizing Risks

1.5.1 "Incremental" Versus "Background" Risks

This way of categorizing risks allows us to estimate the casualty rate that is due to a particular activity, exposure, or occupation and compare that risk with the normal background risk shared by everyone.

For instance, records may tell us that firefighters develop emphysema later in life and at a higher rate than non-firefighters. We can easily compare the incidence of emphysema between firefighters and ordinary (non-firefighter) citizens. But it is important to realize that while firefighters have a higher incidence than the public, not all of their risk is due to their occupational exposure. This is because firefighters are ALSO members of the public and carry the public *background* risk as well as the additional *(incremental)* risk from their occupational exposure. The firefighters' *total risk* is therefore their *background risk + the incremental risk* from their job. The relationship is calculated as:

$$\textbf{Total risk} = \textbf{background risk} + \textbf{incremental risk}$$

Figure 1.3 presents an opportunity to practice this type of calculation using risks for chemical workers.

Analysis challenge:

Assume hypothetically that workers in the Xylol industry have on average 3 chances in a thousand of developing a certain form of cancer in their <u>lifetime</u> (3×10^{-3}). Also assume that in the entire population an average of **two** citizens die from this cancer for every ten thousand population (2.0×10^{-4}). Estimate the workers' risk that is due **just** to workplace exposure (their **incremental risk**).

Fig. 1.3 Finding the incremental risk

Solution to Fig. 1.3 The **total risk** for these chemical workers (3×10^{-3}) is the sum of their ***background risk*** as citizens (2.0×10^{-4}), plus their ***incremental risk*** (due to their work exposure). By switching terms, we can find the incremental risk:

$$\textbf{Total R} = \textbf{Background R} + \textbf{Incremental R}$$

and **Incremental R** = **Total R** − **Background R**

so **Incremental R** $= \left(30 \times 10^{-4}\right) - \left(2 \times 10^{-4}\right) = \left(28 \times 10^{-4}\right)$

The ***incremental risk*** due to their exposure is the difference between ***background risk*** and the ***total risk.*** Note that subtracting the ***background risk*** from the ***total risk*** is easier to do if we use the same exponent; thus, 3 chances in a thousand (3×10^{-3}) can also be read as 30 chances per 10,000, or (30×10^{-4}).

The ***incremental risk*** in this example is therefore 28 per 10,000 (or 2.8 per 1,000, or 0.28%), and the interpretation would read: "Workers can be described as being exposed to *fifteen times the risk* borne by any member of the general population (30 per 10,000 vs. 2 per 10,000), BUT their specific <u>risk due to the workplace</u> is *14 times larger* than the risk borne by a member of the general public: (28 per 10,000 vs. 2 per 10,000). This example demonstrates the importance of clear and precise wording.

1.5.2 "Catastrophic" Versus "Chronic" Risk

Catastrophic risk is the calculated probability of a harmful or injurious event happening, whether or not it has happened so far. Examples include the possibility of a train derailment at a particular piece of track, or the probability that a nuclear reactor containment vessel will rupture, or the likelihood that a set of valves will fail and result in a toxic release. This will be the focus of Chap. 2 of this book.

Chronic risk, on the other hand, is the estimated probability of long-term adverse effects of prolonged (often lifetime) exposure to potentially harmful agents at low doses. Typically, these exposures would be through airborne routes, ingested in food or water or pharmaceuticals, or through the skin. The outcome could be death, disease, or reproductive (congenital or fertility) abnormalities. An example might be the risk of a petroleum refinery worker dying of cancer if they were breathing air containing benzene at a concentration of 4 mg/M^3 during each 8-h working day for 22 years. This will be the focus of Chap. 3 of this handbook.

1.6   Risk and Hazard

The terms **risk** and **hazard** are often used interchangeably. Both terms can vary in their application and common use, but to avoid confusion, we shall define **risk** (quantitatively) as the *product of the probability of the hazard occurring, and the magnitude of its occurrence if it does occur*: ($\mathbf{R = P \times M}$).

A **hazard** can be described as the *potential* of a substance or device to produce harm. "Hazard" is sometimes considered a descriptive term, synonymous with "*danger*," and referring just to the intrinsic capability of the place, process, or material to cause harm, that is, the *source* of the risk. In Chap. 6, we shall discuss the models advanced by Peter Sandman et al., "redefining" risk to take into account some of the nontechnical parameters and heuristics that citizens use to assess risk. Note that the term "hazard" is used in risk perception discussions in a slightly different way.

1.7   Relative Risk and Odds Ratio

1.7.1   Relative Risk

No "toolbox" for measuring "risk" would be complete without discussing two of the most common measures used to compare risk: relative risk and its estimate, the odds ratio.

The relative risk (RR), also called the risk ratio, appears in field calculations, research projects, randomized controlled trials, and during epidemiological investigations of outbreaks. The RR tells you how much more likely you are to experience the outcome (whether good or bad) if you are in one exposure group compared to the other. This method requires the **true incidence rates** for both the "exposure" group (I_e) and for the comparison "nonexposed" group (I_o) and is calculated using the relationship:

$$RR = \frac{I_e}{I_0} \text{ or } \frac{\text{incidence rate among exposed}}{\text{incidence rate among nonexposed}}$$

Each incidence rate has, as the denominator, the number of healthy people originally in the group before the outcome occurred, with the numerator the count of those individuals who eventually develop the outcome, which may be death, a symptom, improvement, or recovery.

As an example, beginning in 1950, a series of large-scale retrospective and prospective investigations were made by epidemiologists Doll, Hill, Peto, and others into the relationship between cigarette smoking and lung cancer, and follow-up was continued through several decades. The data shown in Fig. 1.4 have been greatly simplified for presentation, but the relationship between rates is accurate. Beginning with the records of 30,000 heavy cigarette smokers, and 60,000 nonsmokers, and with access to the diagnoses of lung cancer in each group, the calculation of the *annual* incidence rate for each group can be completed, leading to the relative risk (Fig. 1.4).

	Number with diagnosis of lung cancer per year	Number without diagnosis of lung cancer per year	All
Smokers ('exposed')	39	29 961	30,000
Non-smokers (non-'exposed')	6	59,994	60,000
All	45	89,955	90,000

Fig. 1.4   Relative risk. Example from Doll and Hill (1950, 1954)

The (annual) incidence rate (I_e) for the exposed group = 39 cases/30,000, or 0.0013 per year

The (annual) incidence rate (I_o) for the nonexposed group = 6 cases/60,000, or 0.0001 per year

$$\mathbf{RR} = \frac{\mathbf{I_e}}{\mathbf{I_0}} = \frac{\mathbf{0.0013}}{\mathbf{0.0001}} = \mathbf{13}$$

Interpreting the relative risk We interpret this by observing that a heavy smoker (as defined in the study) had a risk of lung cancer *thirteen times as high as a nonsmoker*. In this way, the **relative risk** is also a measure of the **strength-of-association** (SOA) between the *independent variable* (here, heavy smoking, categorized as yes/no) and the *dependent variable* (annual incidence of lung cancer diagnosis, categorized as yes/no).

A relative risk close to ONE occurs when the incidence rates are similar in both groups, and this can be interpreted as "no association" or "no relationship" between the variables.

Where RR is clearly NOT = 1.0 (i.e., either >1.0 or <1.0), we consider that there does appear to be a relationship between the variables. The "direction" of that relationship depends upon the RR value and also how the table is drawn. In Fig. 1.4, clearly the incidence of lung cancer among smokers was greater than the incidence among nonsmokers.

As a second example, we consider the hypothetical results of a study into the relationship between diarrheal diseases and feeding practices. Healthy newborn infants (n = 120) fed on breast milk were compared with healthy newborn infants (n = 142) fed on bottled formulae in a developing country (Fig. 1.5). They were observed to record incidence of diarrheal disease.

This allows us to calculate the true incidence as: **RR = (7/120)/(18/142) = 0.4602**

As 0.4602 is not close to 1.00, we can claim a relationship, and a closer look at the incidence rates reveals that the breast-fed babies were LESS at risk (7/120 = 0.053) than the bottle-fed babies (18/142 = 0.127).

	Diarrhoeal disease	No diarrhoeal disease	All
Infants breast-fed	7	113	120
Infants bottle-fed	18	124	142
All	25	237	262

Fig. 1.5 Relative risk (enteric disease in infants)

	Diarrhoeal disease	No diarrhoeal disease	All
Infants bottle-fed	18	124	142
Infants breast-fed	7	113	120
All	25	237	262

Fig. 1.6 Relative risk (enteric disease in infants) inverted

Interpreting the results of Fig. 1.5, we can say the risk of diarrheal disease among breast-fed babies is less than half (46%) of the risk among bottle-fed babies. But in a case such as this, it may be more helpful to the reader to display the table such that the greater risk appears as the numerator of the RR. In Fig. 1.6, the same data are displayed but with the rows switched, showing the increased risk as the upper value. This makes it easier to identify and assess how *much* riskier it was than the other exposure. In either case, we must always clearly explain to the reader the "direction" of the effect: *Which* exposure was harmful? *Which* was protective?

In this configuration,

$$\mathbf{RR} = \left(18/142\right)/\left(7/120\right) = \mathbf{2.173}$$

This result clearly identifies bottle-feeding as the more harmful exposure (increased risk), and breast-feeding as the "protective" exposure.

The tables do not have to be redrawn; we can simply invert the initial RR value (divide 1 by the RR giving 1/0.4602 = 2.173) and summarize as follows: *"Bottle-fed babies show more than two times increased risk of diarrhea compared to breast-fed babies."*

How do we know if we have the true incidence data? The previous examples clearly start with 142 healthy-exposed and 120 healthy-comparison groups (denominators). We also have the numerators (18 and 7) as the reported cases of illness or injury that subsequently occurred from each group, giving us valid incidence rates.

Compare the investigation (Fig. 1.7) into a suspected chemical injury. It compares 33 injured workers ("cases") with 28 workers who are not injured ("controls").

The investigators began with the *outcomes* (the ill and the non-ill) and went backward to compare exposures for each. This is commonly called a case-control study, and for these, the relative risk is usually not valid because incidence rates are not available.

Why are there no incidence rates? The totals on the right are not denominators of any incidence rate. No group of 36 ever existed of whom 28 became ill. That is not how the study was conducted.

The initial groups were 33 "ill" and 28 "non-ill" ("*cases*" and "*controls*," respectively), and these are the denominators of "exposure rates," calculated vertically. For example, 28/36 is meaningless. But we can confirm that of 33 cases, 28 (85%) had been exposed, and only 8 of 28 (29%) of the controls had been exposed. These are not incidence rates but exposure rates.

	Ill/ Injured ('cases')	Not ill/injured ('controls')	tot
Exposed to chemical	28	8	36
Non-exposed to chemical	5	20	25
Tot	33	28	61

Fig. 1.7 Odds ratio contingency table

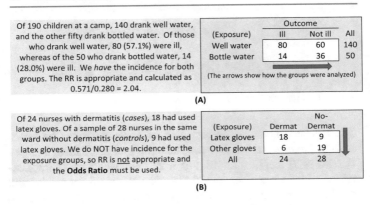

Of 190 children at a camp, 140 drank well water, and the other fifty drank bottled water. Of those who drank well water, 80 (57.1%) were ill, whereas of the 50 who drank bottled water, 14 (28.0%) were ill. We *have* the incidence for both groups. The RR is appropriate and calculated as 0.571/0.280 = 2.04.

| | Outcome | | |
(Exposure)	Ill	Not ill	All
Well water	80	60	140
Bottle water	14	36	50

(The arrows show how the groups were analyzed)

(A)

Of 24 nurses with dermatitis (*cases*), 18 had used latex gloves. Of a sample of 28 nurses in the same ward without dermatitis (*controls*), 9 had used latex gloves. We do NOT have incidence for the exposure groups, so RR is <u>not</u> appropriate and the **Odds Ratio** must be used.

| | | No- |
(Exposure)	Dermat	Dermat
Latex gloves	18	9
Other gloves	6	19
All	24	28

(B)

Fig. 1.8 (**a**) An example requiring use of the true relative risk. (**b**) An example requiring use of the odds ratio

Without the true incidence rates, we cannot calculate the RR, but a substitute exists for calculating risk under these conditions: the *odds ratio* (*OR*). This is almost always used for case-control studies, or any other study where incidence rates are unavailable. Figure 1.8a, b illustrates another pair of situations where a decision is needed: RR or OR?

1.7.2   The Odds Ratio (OR)

The description of the study in Fig. 1.8a indicates that incidence rates *are* available, and therefore, the RR is appropriate.

Now compare with the design of Fig. 1.8b. Here, the incidence rates are unavailable; we only have a small sample of the ill and non-ill people. The odds ratio (OR) is used here as an estimate for the RR. In this instance, instead of starting with the total in each *exposure* group, we gather a sample of *"ill"* persons, and another sample of *"non-ill"* persons, with no assumption that they are present in a representative ratio.

Let's take a closer look at that outbreak of dermatitis among hospital staff. Twenty-four staff members have a skin rash (the "cases"), and latex gloves are suspected to be the cause (Fig. 1.9). A sample (n = 28) of staff in the same area of the hospital who

	Dermatitis		No dermatitis		
Latex gloves	**18**	(a)	(b)	9	27
Other gloves	6	(c)	(d)	**19**	25
	24			28	52

Fig. 1.9 Odds ratio applied to investigation

don't have dermatitis are recruited as a comparison (the "controls"). Note, we clearly do *not* have the incidence data, because we did NOT begin with 18 + 9 = 27 "exposed," or 25 "nonexposed." Instead, we began with 24 "cases" and 28 "controls" and worked backward to record exposure. The cells in the 2 × 2 table are labelled "a" to "d" as shown.

The odds ratio is calculated as $(\mathbf{a} \times \mathbf{d})/(\mathbf{b} \times \mathbf{c}) = (18 \times 19)/(9 \times 6) = 342/54 = 6.33$.

Interpreting the odds ratio An odds ratio very close to 1.0 means "no association," while any other value (either >1.0 or <1.0) suggests that a relationship of some kind exists. (We are not testing for statistical significance, just the *strength of the association*.)

The relationship suggested by the OR of 6.33 requires a *direction*, and regardless of how the table has been constructed, this is easy if we first identify the *"dominant pair"* of cells. To illustrate, if $(\mathbf{a} \times \mathbf{d})/(\mathbf{b} \times \mathbf{c})$ is >1 (as in Fig. 1.9), the $(\mathbf{a} \times \mathbf{d})$ product $(18 \times 19 = 342)$ is greater than the $(\mathbf{b} \times \mathbf{c})$ product $(6 \times 9 = 54)$, so the "a" and "d" cells when multiplied together become the *"dominant" pair* (shown bold in the table). Now look at the labels of those cells: The "a" cell is in the row *"Latex"* and the column *"Dermatitis."* The "d" is in the row *"other"* and the column *"no dermatitis."* Hence, latex glove use is more associated with dermatitis, while "other" gloves are more associated with "no dermatitis." We now have the direction of the relationship.

What about the *strength* of the relationship? That is found in the numerical value of the OR itself. In this case, this is 6.33. We can summarize by stating: *There was a relationship between the type of gloves worn and having dermatitis. The dermatitis people were more than 6 times as likely to have used latex gloves than "other" gloves*. Note that because the odds ratio is used precisely when we do not have the incidence data, it becomes technically invalid to claim that: *"if you used latex gloves, you had 6.33 times the risk of dermatitis."* For this reason, interpret from the perspective of the outcome: *"If you had dermatitis, you were 6.33 times as likely to have been in the latex group."*

The odds ratio therefore yields <u>three</u> useful pieces of information, similar to the RR.

1. **That an association exists** between these variables [when the OR is $\neq$ 1].
2. **Direction of the association** from "dominant pairs"; in this example, cases were more likely to have used latex than "other" gloves [because $(\mathbf{a} \times \mathbf{d}) > (\mathbf{b} \times \mathbf{c})$].
3. **The strength of that association** was the OR value itself. In this example, a person who had dermatitis was 6.33 times more likely to have used latex gloves.

The situation in Fig. 1.10 is a little different. The OR is 0.55, and the dominant pair is therefore $(\mathbf{b} \times \mathbf{c})$. The "**b**" cell clearly links *"Exposed"* (row) and *"Well"* (column). The "**c**" cell links *"not exposed"* (row) and *"Ill"* (column), giving the clear direction of the association: This exposure is *protective*, for instance, a therapeutic treatment or antibacterial agent. It is important to always take the direction from the dominant <u>pair</u> of cells, never a single cell value. For example, the <u>largest single cell value</u> is (**a**), but the $(\mathbf{a} \times \mathbf{d})$ product is NOT dominant.

The **strength of the association**[1] (SOA) is the value of the odds ratio itself, and in Fig. 1.10, it is less than 1.0:

	Ill.			Well	
Exposed	130	(a)	(b)	**105**	235
Not exposed	**90**	(c)	(d)	40	130
	220			145	365

Fig. 1.10   When OR is less than 1.0

---

[1]The terms "association" and "relationship" are interchangeable throughout.

$$OR = (a \times d) / (b \times c) = (130 \times 40) / (105 \times 90) = 0.55$$

This can be interpreted by saying "... *the exposure was protective; an ill person was only half (55 percent as likely) to have been exposed compared to a non-ill person.*" Conversely, a more useful and intuitive way to summarize the data may be to focus on the increased probability of exposure among the well group. We do this either by inverting the table or inverting the OR value (1/0.552 = 1.82), in which case we can explain: "*...that a well person was 1.82 times (almost twice) as likely to have been exposed to (X) compared to an ill person. The exposure was therefore protective.*" (No implication can be made at this point concerning statistical significance, which has not been tested for in this example.)

1.7.3   Weakness in the Relative Risk and Odds Ratio

The RR and OR are *ratios*. In other words, the outcome depends on the ratios between the cell values, not their absolute values. As such, they are unreliable when the cell values are small. When calculated by statistical software, a **confidence interval (CI)** is typically included. If the range of the confidence interval includes 1.0, we know we will not be able to claim statistical significance even though a test of significance (such as the chi-square test) has not yet been carried out.

For example, inspect the three tables displayed in Fig. 1.11. Although the cell values are all different, the ratios between cells, calculated by cross-multiplying $(a \times d)/(b \times c)$ are the same. Each

A	Ill	well	Tot
Exp	4	5	9
N/exp	6	10	16
	10	15	25
OR	1.333		
CI 95%	0.254 – 7.007		

B	ill	well	Tot
Exp	40	50	23
N/exp	60	100	12
	30	5	35
OR	1.333		
CL 95%	0.789 – 2.253		

C	Ill	well	Tot
Exp	400	500	180
N/exp	600	1000	320
	200	300	500
OR	1.333		
CI 95%	1.129 – 1.574		

Fig. 1.11 Confidence interval decreases as *N* increases

table produces the same relatively low OR of 1.33, but the confidence intervals are different due to the actual numbers present. The 95% confidence interval is shown for each table.

Tables A and B with the small cell values *include* 1.0 in the confidence interval and would be excluded from the possibility of statistical significance. Table C, with the larger cell values, does not include 1.0 in the confidence interval and could be a candidate for statistical significance if this was tested. (These examples cite the OR, but the same interpretation is appropriate with the RR).

1.7.4   Understanding Confidence Intervals (CI)

Most computations of statistical parameters (e.g., mean, regression coefficient, odds ratio, etc.), typically include a confidence interval around that value. As an example, the arithmetic mean of a sample of values taken at random from a large, normally distributed population might be expressed as: $\bar{x} = 4.200$ (CI $_{95\%}$: 3.000, 5.400). This is interpreted as follows: *The population mean is unknown. However, the mean of the sample* $(\bar{x})$ *is 4.200, and we can be 95% confident that the population mean lies between 3.000 and 5.400.* (These are also called *confidence limits*).

In the case of the confidence interval around an odds ratio, recall that an OR of 1.00 indicates *no relationship* between exposure and outcome. Therefore, if the 95% confidence limits *include* 1.00 in the range, the "no-relationship" scenario is within the 95% probability range for the odds ratio. This would disqualify the relationship from being *statistically significant* at any level. In Fig. 1.11, A and B would be disqualified, but C might still be found to be statistically significant if tested.

1.8 Case Study 1: Driving or Flying?

Case Study #1: Driving or Flying?

How do driving and flying really compare? Assume you need to take a trip to a conference 200 km away; is it safer to fly or drive? For driving, the latest figures from Transport Canada show 4.9 deaths in 1 billion kilometers driven. That's 4.9×10^{-9} deaths per km driven.

However, aircraft travel is relatively uneventful once you reach cruising altitude: close to 90% of accidents happen around takeoff and landing regardless of the distance flown (Fig. 1.12). This means that measuring the risk per kilometer is not a valid measure for air travel.

Safety experts use passenger boardings (PBs) as a measure of a flight, almost disregarding the actual distance. The latest figures (Barnett 2020) show that the risk of fatality globally is one death per 7.9 million PBs (down from one death per 2.7 million PBs during 1998–2007). On a per-PB basis, the fatality risk is $1/(7.9 \times 10^6)$ or 1.26×10^{-7} (deaths per 1 PB).

We can construct a chart (Fig. 1.13) with distance (km) as the x axis and death risk as the y axis. A flat line at 1.26×10^{-7} deaths is a rough representation of the fatality risk for a single flight, regardless of actual distance traveled. The analysis then asks: *"How many kilometers would need to be driven to reach that same risk?"*

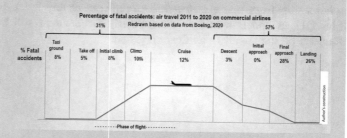

Fig. 1.12 Risks of fatality on phases of commercial flight

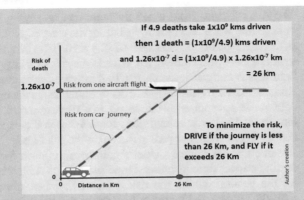

Fig. 1.13 Comparison between flight risk and driving risk

The solution, surprisingly, is only about 26 km!

About 15 years ago, the same calculation showed close to 350 km, but since then, flying has become safer and the roads and driving have become more dangerous.

Of course, this comparison is oversimplified and doesn't take into account the numerous other variables of driving (driver's age, road condition, car condition, traffic, even the drive to the airport) or of flying (whether large turbofan/jet or smaller commuter and charter flights). But the method is instructive.

Practice Exercises: Chapter 1

[A] The risk of death from liver cancer for chemical workers is 1 in 250,000, and the risk of death from the same form of cancer for an average person in the population is 1 in 5 million.

  (a) What is the incremental risk for a typical worker of this type?

  (b) What is the background risk for a typical worker of this type?

  (c) What is the total risk for a typical worker of this type?

[B] The risk of death from a certain brain cancer among high-voltage line workers is (1×10^{-5}), but for the general public, the risk of this type of cancer is 2.4E-07 (2.4×10^{-7}). (a) What is the total risk for a line worker? (b) What is the risk of this type of brain cancer that can be assumed to be due to the work exposure?

[C] Roofers are faced with the risk of death from falls. In the region, 3600 roofers are currently employed. In a 10-year period, there have been 18 deaths among this group. What is the expected annual risk of death per person due to on-the-job falls?

[D] Assume that the annual risk of death from a certain form of thyroid cancer among radioisotope workers is 2.5E-05, but for the general public, the risk from this type of cancer is 5.0E-08. What is the incremental risk for this worker?

[E] The annual risk of a certain reproductive disorder among petrochemical workers is 2.1E-04, but for the general public, the risk is 5.8E-07. What risk can be attributed to this work exposure?

[F] Workers manufacturing brake pads are found to have a lifetime rate of mesothelioma close to 4 per 100,000. The same disease appears in the general population at a rate of 8 per one million. What is (a) the background risk, (b) the risk attributed to the work, and (c) the total risk?

[G] New fall-arrest equipment. A new device for arresting the fall of a worker is claimed to safely absorb more of the momentum, thus preventing injury. Approximately 400 workers are equipped with the new equipment, while another 400 workers use the existing equipment for 6 months. The outcome measure for each fall was: *When it was used, did it result in a musculoskeletal injury requiring orthopedic treatment?* There were 44 falls during the test period. The data are presented below. How would you summarize the exposure risk of the two groups? And what is your recommendation about whether the new equipment reduces the risk of injury?

	Requiring orthopedic treatment	Not requiring orthopedic treatment	Totals
Existing equipment	9	10	19
New equipment	14	11	25
	23	21	44

[H] Eight hundred pilots regularly fly helicopters in the province. Sixteen work-related deaths were recorded in a 4-year period. (a) What is the annual work-related risk of death per helicopter pilot? (b) Assume 2,560 fixed-wing pilots are licensed and currently flying in Ontario. In a 10-year study, there had been 64 deaths. Which type of flying has the greater annual risk of death for the pilot, helicopter, or fixed-wing?

[I] The hypothetical annual incidence of a certain form of bladder cancer in the Ontario population is 4 in 100,000. That same form of cancer has been measured in fabric dye manufacturing workers, at a rate of 2 per 10,000. Estimate the incidence of bladder cancer that can be attributed to working in the fabric dye industry.

[J] Discovery of MRSA infection among 11 patients following their surgery has led to the hypothesis that one of the surgical team was carrying MRSA. Personnel are being swabbed and the results will be known in 48 h. You are calculating a patient's risk of being infected (M+) for five members of the surgical team using odds ratios. All the surgeries completed in a 14-day period are listed and the presence or absence of that surgical team member during surgery is shown for the 11 MRSA+ patients and the 42 non-MRSA patients. Calculate the risk of infection for each team member in terms of the increase risk a patient will be infected if that team member was present, compared to when they are not present.

	Anaesth #1			Nurse #2			Nurse #3			Surgeon #4			Surgeon #5		
	M+	M–	Tot	M+	M–	Tot	M+	M–	Tot	M+	M–	Tot	M+	M–	Tot
Present	1	4	5	2	8	10	7	28	35	4	16	20	10	14	24
Absent	10	38	48	9	34	43	4	14	18	7	26	33	1	28	29
Tot	11	42	53	11	42	53	11	42	53	11	42	53	11	42	53

Question	Solution and comments
A	(a) 3.8×10^{-6} (b) 2.0×10^{-7} (c) 4.0×10^{-6}
B	(a) 1.0×10^{-5} (b) 9.76×10^{-6}
C	1.8 deaths/3600 workers per year $= 5.0 \times 10^{-4}$
D	$2.5 \times 10^{-5} - 5.0 \times 10^{-8} = 2.495 \times 10^{-5}$
E	2.094×10^{-4}
F	(a) 8×10^{-6} (b) 3.2×10^{-5} (c) 4×10^{-5}
G	OR:0.71: (Inverting:1.41) Falls requiring orthopedic treatment were 41% more likely to have involved the new safety equipment. The numbers are very small, but the suggestion is that the new safety equipment does not provide better protection against more severe injury.
H	H/pilot risk: 0.0050; Fixed-w/pilot risk 0.0025; H/p twice the risk of death
I	$2/10,000 - 4/100,000 = 1.60 \times 10^{-4}$
J	An MRSA+ patient was 20 times more likely to have had surgeon #5 operating.

References

Barnett A. Aviation safety: a whole new world? Transp Sci. 2020. https://doi.org/10.1287/trsc.2019.0937. https://pubsonline.informs.org/doi/10.1287/trsc.2019.0937

Boeing Corporation. Statistical summary. 2020. https://www.skybrary.aero/sites/default/files/bookshelf/32664.pdf. Accessed on 5 Nov 2022.

Doll R, Hill AB. Smoking and carcinoma of the lung. Br Med J. 1950;2:739–48. https://www.ncbi.nlm.nih.gov/pmc/articles/PMC2038856/pdf/brmedj03566-0003.pdf. Accessed 17 Nov 2022.

Doll R, Hill AB. The mortality of doctors in relation to their smoking habits. Br Med J. 1954;6:1451–5. https://www.ncbi.nlm.nih.gov/pmc/articles/PMC2085438/pdf/brmedj03396-0011.pdf. Accessed 17 Oct 2017.

International Standards Organization. ISO 31,000. Risk management. Geneva: ISO Central Secretariat; 2019.

National Research Council (NRC). Science and decisions: advancing risk assessment. Washington, DC: The National Academies Press; 2009.

Probabilistic Risk Assessment

2

Abstract

This chapter predicts and measures accidents, incidents, and catastrophic events that may or may not occur in the future. The probabilistic risk assessment (PRA) form of analysis has applications across a very wide landscape of sectors, technologies, and systems in fields as far apart as oil drilling, surgery, management, and aerospace engineering. Probabilistic risk assessment deploys several analytical methods, including formulae, contingency tables, Venn diagrams, and especially probabilistic trees. These resemble fault trees, but specific probability values are inserted at every intersection.

For those who are unfamiliar with the language, calculation, and interpretation of probabilities, or have become a little rusty in these techniques, a probability refresher course has been included at the beginning of this chapter.

2.1 Risk and Uncertainty

Predicting the probabilities of outcomes that have not yet occurred invariably requires the risk analyst to deal with uncertainty. This crystal ball comes with a toolbox of well-tried methods and instruments.

The original version of the chapter has been revised. A correction to this chapter can be found at https://doi.org/10.1007/978-3-031-28905-7_7

© The Author(s), under exclusive license to Springer Nature
Switzerland AG 2023, Corrected Publication 2024
E. Liberda, T. Sly, *Assessment and Communication of Risk*,
https://doi.org/10.1007/978-3-031-28905-7_2

The language of risk is the language of probability, where every result exists on a scale between zero and one. We all use percentages in predicting a wide range of outcomes, while probabilities are simply percentages with the decimal point moved two places to the left, such that 100% is a probability of 1.0; 50% is a probability of 0.5; 8% is a probability of 0.08 and so on.

Our language often reflects a fascination with uncertainty and probability: *"What are the odds...?" "What are the chances...?" "I'll bet you..." "A snowball's chance in hell!" "Beat the odds!" "Toss of the dice" "A crap shoot"* etc. However, in assessing risks to health and safety, we need more objective estimates such as numerical probabilities at each stage, and just as important, a measure of confidence around those estimates. When *risk estimates* are reported in the press, the uncertainty surrounding those estimates is often unreported, and when it is quoted, it remains unexplained, which is unfortunate as the picture is not as complete as it should be.

Figure 2.1 shows a selection of assessed risks accompanied by an uncertainty factor for each (extract from Wilson and Crouch 1987). As we shall discuss later in this text, some of the uncertainty is *systematic*, arising from the methods of risk assessment being used (including the overuse of worst-case scenarios), while other uncertainties include genuine *variability* from the interplay of unknown variables. In this example, the uncertainty ranges from 5% (.05) to a factor of 10, giving a range of 200 times more variability in assessing drinking water containing TCE than assessing the risk of death from home accidents.

Probability theory is the foundation of this field, and we will explore it only to the extent necessary for us to appreciate and use the tools it provides. Without these tools, our work will be subjective and lack credibility.

The calculations of probability originated in the games of chance: dice, cards, roulette, etc., in European casinos. These

	Annual Risk	Uncertainty
Home accidents (deaths)	1.1×10^{-4}	5%
Motor vehicle accident (total deaths)	2.4×10^{-4}	10%
Motor vehicle accident (pedestrian deaths)	4.2×10^{-5}	10%
Cigarette smoking (1 pack/day)	3.6×10^{-3}	factor of 3
Peanut butter (4 teaspoons/day)	8.0×10^{-6}	factor of 3
Drinking water at regulatory limit of TCE	2.0×10^{-9}	factor of 10

Fig. 2.1 Examples of risks and their uncertainty

games were played for centuries, but it was not until aristocratic gamblers asked mathematicians such as Pierre de Fermat and Blaise Pascal for help that the branch of mathematics now known as *probability theory* arose.

2.2   **Modeling Probability**

When we toss a coin, we cannot know the outcome in advance. So what can we predict? We know that the outcome will be either heads or tails and nothing else. And because the coin appears to be balanced, each of these outcomes has the probability of showing 50% of the time or with a probability of 0.5. Some of the essential characteristics of a probability value are shown in Fig. 2.2.

Sometimes you may be seeking the probability of a combination of outcomes that satisfies a certain requirement or "argument." Figures 2.3 and 2.4 emphasize the importance of *clearly* specifying what constitutes an outcome or "event." Let's be clear about what we mean by an *event:* An **"event"** is any individual outcome or set of outcomes that is a subset of the sample space.

If the "event" was specified as *"exactly two heads in all possible sequences of four throws of a coin,"* we need to identify that there are exactly sixteen possible outcomes in throwing four coins, as shown in Fig. 2.3. Of these sixteen, only six meet the requirements of the stated event. The set of outcomes that satisfies the *"exactly two heads"* argument is {HHTT, HTHT, HTTH, THHT, THTH, and TTHH}. Thus, SIX out of a possible sixteen gives us 6/16 or 0.375, and we could state, using the conventional format,

$$\mathbf{P}\left(\textit{exactly two heads in four throws}\right) = \mathbf{0.375}.$$

---

1.   A probability is a number between 0 and 1.
2.   The list of all possible outcomes is known as the "sample space", [S].
3.   The probabilities of each of the outcomes in the sample space must add to 1.00

Example: Toss a coin once. Only two outcomes are possible,
-    the sample space [S] is {H, T} and the [P] of each outcome is 0.5
Example: Stick a pin in a table of single, random, digits: Ten outcomes are possible,
.   the sample space [S] is {0,1,2,3,4,5,6,7,8,9}, and the [P] of each is 0.1

Fig. 2.2   Basic characteristics of a probability

Toss a coin four times. Record the order of the H and T for each set.																
A single outcome will be one of the following 16 combinations that comprise the sample space																
	1	2	3	4	5	6	7	8	9	10	11	12	13	14	15	16	
	H	H	H	H	T	H	H	H	T	T	T	H	T	T	T	T	
	H	H	H	T	H	H	T	T	H	H	T	T	H	T	T	T	
	H	H	T	H	H	T	H	T	H	T	H	T	T	H	T	T	
	H	T	H	H	H	T	T	H	T	H	H	T	T	T	H	T	
The P for each of the outcomes is 1/16 or 0.0625																

Fig. 2.3 Counting probabilities

Toss a coin four times. Record the number times that H appears						
Event	4 Heads	3 Heads	2 Heads	1 Head	0 Heads	Total
Fraction Probability	1/16 or 0.0625	4/16 or 0.2500	6/16 0.3750	4/16 0.2500	1/16 0.0625	16/16 1.0000

Fig. 2.4 Possible number of heads in four throws of a coin

Figure 2.4 lists the probabilities of all possible outcomes where the number of "heads" is specified. (This type of analysis is carried out intensively in casinos to compute the true chance of winning, which is always higher than the actual payout that is made should you win).

In probability models, all events have probabilities. We can write the probability of event A as **P(A)**. Figure 2.5 demonstrates an example using playing cards.

Challenge: From a normal pack of 52 playing cards, find the probability of drawing at random: a) Any diamond;   b) Any picture card;   c) Any king;   d) The ace of spades;   e) Any red card;
[Answers: a) 0.2500   b) 0.2308   c) 0.0769   d) 0.0192   e) 0.5000]

Fig. 2.5 Counting probabilities (single cards)

2.3   **Addition of Probabilities**

So far, we have discussed coin-tossing, dice, and cards, and while these seem remote from health and safety issues, you will see the close connection when we begin to calculate probabilities of human health risks using these same techniques.

Many of the probabilities in practical situations are not equal, although they still add to 1.0. Look at Fig. 2.6 and ask: *"What is the probability of selecting at random a red or an orange candy?"* We accept that *either* a red *or* an orange piece will meet the requirement so that the **P(red or orange)** will be greater than either **P(red)** or **P(orange)** separately.

In the world of probability, the word "*or*" denotes the simple addition of two separate probabilities, but only where the events are **mutually exclusive.** What does that mean? When two or more items are mutually exclusive, selecting just one of them has excluded any other from being selected. In the example of the candies, each is a single color; no one could select a candy that is both red *and* orange. So the colors are mutually exclusive and simple addition is appropriate. A single candy can ONLY be red or yellow or green, etc. This gives us the special rule for adding probabilities where the outcomes are quite distinct:

> *Simple addition rule* $($ *only for mutually exclusive events* $)$:
> $P(A \, or \, B) = P(A) + P(B)$

Applying the addition rule to the Venn diagram in Fig. 2.7, we simply **add** the probabilities of red (0.2) and orange (0.1).

$$P(\text{red or orange}) = 0.3$$

You have a large bag of mixed, coloured candy pieces. The sample space is {brown, green, orange, red, tan, and yellow}. The probability of each outcome has been obtained from the manufacturer's record of all candies made. This we learn to be 30% brown, 20% red, and so on. Here is the complete set (the "sample space")

	brown	red	yellow	green	orange	tan	Total probability
probabilities:	0.3	0.2	0.2	0.1	0.1	0.1	1.00

Fig. 2.6 Adding probabilities with variable values

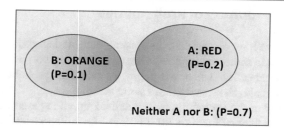

Fig. 2.7 Adding probabilities using a Venn diagram

Don't forget that *everything within* the box becomes the *sample space*, and it all must add to 1.0. In this example,

$$\mathbf{P}\big(\textbf{neither red nor orange}\big) = \mathbf{0.7}$$

Figure 2.8 depicts another mutually exclusive situation involving random digits. Here, we set event (A) to be any odd digit; there are five odd digits, so P(A) = 5/10 or 0.5. Event (B) is a multiple of 4, and by convention, this includes zero, so 0, 4, and 8 are included. P(B) = three out of ten, 3/10 or 0.3. What then is **P(AorB)**? First, decide if P(A) and P(B) are mutually exclusive: Does any value appear to satisfy <u>both</u> A and B? In this case, no values for A can be included in B. They are *mutually exclusive*.

The solution **P(AorB)** includes the five outcomes in A plus the three in B. Eight digits satisfy this requirement: 1, 3, 5, 7, 9, and 0, 4, 8. So **P(AorB)** = **P(A)** + **P(B)** = 5 + 3 = 8, and 8/10 = 0.8.

How, then, does the addition of two or more probabilities change when the events are *not* mutually exclusive? Figure 2.9 is an instance where an outcome can belong to both categories. If a selection can satisfy *both* requirements, then the simple addition does *not* apply. For instance (Fig. 2.9), in throwing a six-sided die, we can define event (A) as rolling an even number, and event (B) is defined as rolling a number less than or equal to 4. (A) has three outcomes: 2, 4, 6, so P(A) = 3/6 = 0.5, while (B) has four outcomes: 1, 2, 3, 4, so P(B) = 4/6 = 0.67.

Event (A) is <u>not</u> mutually exclusive of (B) because two of the outcomes (2 and 4) satisfy <u>both</u> (A) <u>and</u> (B). We now have a new quantity: the zone where <u>both</u> A and B are outcomes. Figure 2.9 illustrates this. The new zone, **A**and**B** (sometimes written A∩B) is

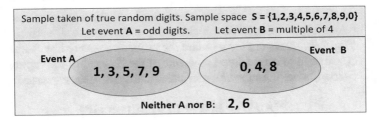

Fig. 2.8 Adding mutually exclusive probabilities

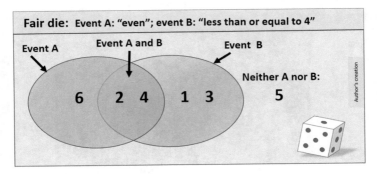

Fig. 2.9 Adding probabilities with common properties

an "overlap" between the two main events, A and B. The word "*and*" takes on a somewhat restrictive nature in probability because *both A and B must be satisfied*. This is unlike the everyday concept of "*and*" which is more inclusive (thus larger than each of the component parts). An important concept here is that because A and B are *not* mutually exclusive, we must use the **general rule for addition** of probabilities to find the probability **P(A or B)**:

General rule for addition of probabilities :
$$P(A\,or\,B) = P(A) + P(B) - P(A\,and\,B)$$

The general rule is actually appropriate for *all* additions. This is because if the two events *are* mutually exclusive (no overlap), then "AandB" will be zero and can be ignored. If A and B are NOT mutually exclusive, as here, then think of the common area ("AandB") as being TWO "thicknesses" of the measured "surface" (Fig. 2.9). To accurately measure the exact area covered by A and B, we need to surgically remove *one* of the "layers" of the central overlap, hence the subtraction included at the end of the expression:

$$\mathbf{P(AorB)} = \mathbf{P(A)} + \mathbf{P(B)} - \mathbf{P(AandB)}$$

Figures 2.10 and 2.11 illustrate the difference between "AorB" (the whole entire shaded area in Fig. 2.10) and "AandB" (the restricted "overlap" in the center of Fig. 2.11).

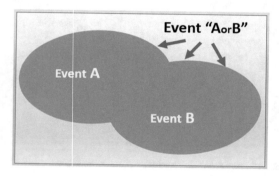

Fig. 2.10 Defining (AorB) or ("A∪B")

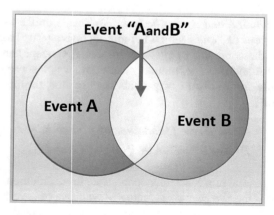

Fig. 2.11 Defining (AandB) or ("A∩B")

2.4 Complementary Events

The *complement* of event A is the sum of everything in the sample space that is NOT event A. We denote the complement of event A by the symbol A^c (Fig. 2.12). The complement of a probability is a very useful concept and is used extensively in both simple and complex calculations about risk.

In the Venn diagram, "complement" refers to the space outside the defined zone but inside the box that represents the complete sample space. Figure 2.13 represents an example of complementary probabilities in calculation.

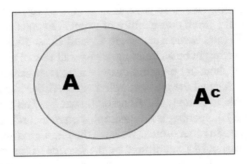

Fig. 2.12 Complementary events

Analysis: Hospital personnel
While auditing annual injury reports for 48 rural hospitals we find that 12 hospitals report no injuries, 12 report one injury, 10 report two injuries, 9 report three injuries, and 5 report four or more injuries during the year. We want to find the probability of a hospital reporting *one or more injuries*. Selecting a report with one or more, is the *complement* of selecting a report with a NO injury. Let 'NO injury' = event A, thus the **complement** of A will be 'one or more'.

$P(A^c) = 1 - P(A) = 1 - 12/48 = 36/48 = (0.75)$

Alternatively, taking longer, we could have calculated the probability by adding
= P(1 injury) + P(2 injuries) + P(3 injuries) + P(4 or more injuries)
= 12/48 + 10/48 + 9/48 + 5/48 = 36/48 = (0.75)

Fig. 2.13 An application of complementary probabilities: Hospitals

Rule for complementary events : If A^c is the
complement of event A, then $P\left(A^c\right) = 1 - P\left(A\right)$
E.g., If $P\left(A\right)$ *was 0.4, then* $P\left(A^c\right) = 1 - 0.4 = 0.6$

2.5   Multiplication of Probabilities

We have already been dealing with multiplying probabilities: Recall **P(A**and**B)**? That was really asking: *"What's the probability of (event A occurring and then followed by event B?)"*

In assessing risk with several factors, we frequently need to find the final (*'joint'*) probability of event A happening, followed by event B (and sometimes event C, and D, and so on). As an example, we might be calculating the overall risk of the following scenario: A chemical pressure vessel exceeds its maximum pressure (event A), and the pressure relief valve ALSO becomes stuck and is unable to remedy the situation (event B). The consequence of (A and B) happening is a catastrophic chemical release into the environment. In other words, we have the *likelihood* (probability) of high pressure (A), multiplied by the *likelihood* (probability) of the pressure relief valve (B) failing when it was most needed. BOTH A and B need to happen to produce the catastrophe, and the probability of that is described as P(A and B).

When multiplying these probabilities, we need to be aware of the importance of the sequence and also whether the events are independent or not. Event B *may* fail at the same rate whether A failed or not, or it is possible that event B was more likely to occur because event A had just happened, but we cannot assume this. It is missing information unless provided.

When we multiply events A and B to reach the joint probability P(A and B), we consider the first event, (A) in our example, to simply "happen," but the probability of the event B is conditional upon the previous event having taken place. We could write this as follows:

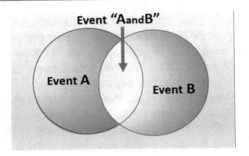

Fig. 2.14   Another look at P(AandB)

$$\mathbf{P}\big(\mathbf{A}\,\text{and}\,\mathbf{B}\big) = \mathbf{P}\big(\mathbf{A}\big) \times \mathbf{P}\big(\mathbf{B},\,\textit{"given that A has already occurred"}\big).$$

This is the general multiplication rule where A and B are two events in sequence.

It is written more conveniently as: $\mathbf{P(AandB)} = \mathbf{P(A)} \times \mathbf{P(B|A)}$,

… and is read as the *"The probability of A, multiplied by the probability of B, <u>given A.</u>"*

If you have trouble "seeing" this conditional argument

$$\mathbf{P}\big(\mathbf{A}\text{and}\mathbf{B}\big) = \mathbf{P}\big(\mathbf{A}\big) \times \mathbf{P}\big(\mathbf{B}\,|\,\mathbf{A}\big)$$

try looking at the comparatively simple illustration of "A and B" in Fig. 2.14. Consider that to measure this central area between (A) and (B), we would first need to select all of (A), and then from that denominator, select what is also part of (B). The result would be the elliptical area shown in the diagram.

2.6   Conditional Probabilities

Rearranging the general multiplication rule, we can also define the conditional probability

$$\mathbf{P}\big(\mathbf{B}\,|\,\mathbf{A}\big)\ \text{is equal to}\ \ \frac{\mathbf{P}\big(\mathbf{A}\text{and}\mathbf{B}\big)}{\mathbf{P}\big(\mathbf{A}\big)}\ \ \dots\ \Big[\text{but only if } \mathbf{P}\big(\mathbf{A}\big) > 0\Big]$$

The reduced area, common to BOTH events, which we have been calling (AandB), is described as a **joint probability** and is obtained as the product of multiplying the P(A), not by P(B) but by the *conditional* P(B|A). This will become especially clear and intuitive when we use probability trees.

Taking a more applied illustration, suppose the average annual risk that a member of the public will need orthopedic surgery is 1 percent. But the average annual risk of needing orthopedic surgery *following a motor vehicle accident (MVA)* is 10 percent. Expressing this in probability terms,

$$\textbf{P}\left(\textbf{orthopaedic surgery}\right) = \textbf{0.01}$$
$$\text{but } \textbf{P}\left(\textbf{orthopaedic surgery}|\textbf{MVA}\right) = \textbf{0.10.}$$

The risk of surgery, *conditional upon a previous MVA*, has increased tenfold. Note that whereas P(orthopedic surgery) used the whole population as a sample space or denominator, P(orthopedic surgery|MVA) uses only those people recently involved in a motor vehicle accident as the sample space. *A conditional probability always has a reduced sample space.*

Without additional information (e.g., the probability of having a motor vehicle accident), at this stage, we cannot find P(MVA and O-S)[1]

It's worth a reminder at this stage that when trying to understand a conditional probability, such as P(A|B), first identify the conditional quantity "B" (to the right of the vertical line), and starting with that as a new denominator or sample space, then identify the quantity that also satisfies the initial term "A."

From Fig. 2.15, we can find the full range of probabilities, including all the conditional probabilities. The first step is calculating the probabilities for V and L.

The denominator is 1400 pts., of whom 532 are found with vascular (V) complications (0.38) and 1260 with lung (L) complications (0.90). Thirty percent have both (V and L) syndromes:

$$(0.30) \times 1400 = 420\,\text{pts.}$$

---

[1]Note that P(AandB) is numerically equivalent to P(BandA), but in this example, MVA comes before O-S, so we have changed the order to P(MVA and O-S).

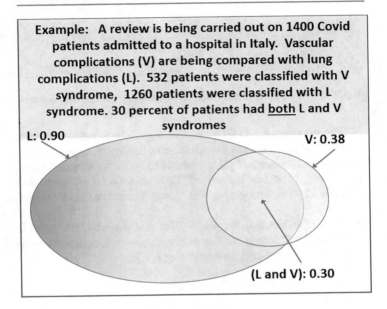

Example: A review is being carried out on 1400 Covid patients admitted to a hospital in Italy. Vascular complications (V) are being compared with lung complications (L). 532 patients were classified with V syndrome, 1260 patients were classified with L syndrome. 30 percent of patients had **both** L and V syndromes

L: 0.90

V: 0.38

(L and V): 0.30

Fig. 2.15 Identifying conditional probabilities

The remaining probabilities can now be calculated:

- **P(V):** 532/1400 = 0.38
- **P(L):** 1260/1400 = 0.90
- **P(VandL):** 420/1400 = 0.30 (given)
- **P(onlyV): P(V) – P(LandV):** 0.38 – 0.30 = 0.08
- **P(V^C): 1 – P(V):** 1 – 0.38 = 0.62
- **P(VorL): P(V) + P(L) – P(LandV):** 0.38 + 0.90 – 0.30 = 0.98
- **P(neitherAnorB): 1 – P(VorL):** 1 – 0.98 = 0.02 (i.e., only 2% of pts had other conditions)
- **P(L|V):** Visually: 0.30/0.38 = 0.7895 [V yields 0.38 and from those, L is 0.30]
- **P(V|L):** This time using the formula: **P(V|L)** = P(VandL)/**P(L)** = 0.30 / 0.90 = 0.333′

These data are obtained from a chemical plant, and describe the status of 66 valves used to control hazardous materials.	Working status	Equipment use			Total
		Low	Moderate	Heavy	
	Satisfactory	10	20	14	44
	Defective	2	8	12	22
	Total	12	28	26	66

Fig. 2.16 Finding probabilities (table): chemical plant valve conditions and use

The contingency table shown in Fig. 2.16 illustrates similar conditional probabilities in an industrial setting. A safety audit of 66 valves in a chemical plant finds 44 to be in "satisfactory" condition and 22 to be "defective." The same 66 valves were also classified by frequency-of-use as "low," "moderate," and "heavy."

- Find **P(defective),** the probability that any random valve is "defective." This is *NOT* conditional, so the denominator is 66 (all the valves), and the numerator is 22:

$$\mathbf{P}(\mathbf{def}) = 22 / 66 = 0.3333$$

- Find **P(def | heavy),** the conditional probability that any random valve is "defective" *given that* it is in "heavy use." The denominator is first *reduced* to just the "heavy use" valves (n = 26), and the numerator is the "defective" count but ONLY from the "heavy use" set (n = 12). Note that when faced with a conditional probability, take the conditional factor (on the right side of the vertical line) first, as the denominator and from that quantity, the other factor as the numerator:

$$\mathbf{P}(\mathbf{def} | \mathbf{heavy}) = 12 / 26 = 0.4615$$

- Find **P(heavy|satisf),** the (*conditional*) probability that a randomly selected valve is in heavy use given that it is satisfactory: **P(heavy|satisf),** = 14/44 = 0.3182, by simply picking the values we need from the table. But the formula produces the same result and should be remembered. Recall:

$$P(A|B) = \frac{P(A \text{ and } B)}{P(B)}$$

$$P(\text{heavy} \mid \text{sat}) = \frac{P(\text{heavy and sat})}{P(\text{sat})} = \frac{14/66}{44/66} = \frac{14}{44} = 0.3182$$

- Find the (*conditional*) probability that a randomly selected valve is defective, given that it is moderately used: easiest to use 8/28 visually from the table, but the formula also obtains the same value:

$$P(\text{def} \mid \text{moderate} - \text{use}) = \frac{P(\text{def and moderate} - \text{use})}{P(\text{moderate use})}$$

$$= \frac{8/66}{28/66} = 8/28 = 0.2857$$

In a second example (Fig. 2.17), a contingency table of data from a hospital-based survey of *Staphylococcus aureus* isolations is shown, with the 272 isolates classified in two ways, as antibiotic resistant (yes/no), and also as pathogenic (yes/no).

P(resistant): 35/272 = 0.12868
P(pathogenic): 29/272 = 0.106618
P(pathogenic and resistant):

Visually, 12 / 272 = 0.04412

Fig. 2.17 *Staph.*
isolations: table display

Pathogenic	Resistant?		
strain?	Yes	No	Total
Yes	12	17	29
No	23	220	243
Total	35	237	272

But using the formula: $\mathbf{P(A and B) = P(A) \times P(B|A)}$ we find the same answer:

$$\begin{aligned} P\left(\text{pathogenic and resistant}\right) &= \mathbf{P\left(path\right)} \times \left(\mathbf{P\left(res|path\right)}\right) \\ &= \left(29/272\right) \times \left(12/29\right) = 12/272 \\ &= 0.04412. \end{aligned}$$

In these tables, each of the sampled items appears *four* times. It will appear in one of the four cells in the central matrix of the table, also in one of the row-totals, one of the column-totals, and of course in the grand total. Each of the cells acquires its value from, or is contingent upon, the cells surrounding it, hence the term *contingency* table. We have already encountered the usefulness of contingency tables in Chap. 1 when we explored the relative risk and odds ratio, and from that work, you should also be able to determine the answer to this question:

In the table (Fig. 2.17), is resistance independent from pathogenicity? Several methods are available. Compare vertical ratios 12:23:35 with 17:220:237. Are they similar? Or compare horizontal ratios 12:17:29 with 23:220:243. But the fastest way is by cross-multiplying (a × d)/(b × c), which gives us 6.7519, clearly a long way from 1.00. These two variables are therefore related (dependent), and as (a × d) is dominant, we can see that pathogenic strains are more likely to be resistant, while nonpathogenic strains are more likely to be nonresistant. In the next section, we will analyze the same data using a probability tree, and the "independence" question will be easier to visualize.

2.7   The Probability Tree

At this stage, we have the ability to add and multiply probabilities and analyze them using formulae, Venn diagrams, and contingency tables. A third method uses a probability tree, and this has the advantage that it can be applied to many (more than two) variables. It resembles a fault tree or a decision tree, but with numerical values placed at every intersection allowing fairly complex analyses to be carried out. Commercial software is also available to calculate the more complex arrays of data.

Probability trees depict events or sequences of events as branches. Our examples show sequences of events from left to right, although there's nothing to prevent the tree from being drawn from top down or right to left, as long as it is correctly labelled. The tree in Fig. 2.18 depicts the same *S. aureus* data as was shown in the table (Fig. 2.17).

Each branch is labelled to indicate which event it represents and the probability of that event's occurrence, and all the branches that emanate from one intersection point must be complete and add to 1.00.

For this first example, at each intersection, both the frequencies (counts) and the probabilities are shown. More often, the probabilities alone are shown. Whether you use frequencies or probabilities depends to some degree upon the data and also upon what

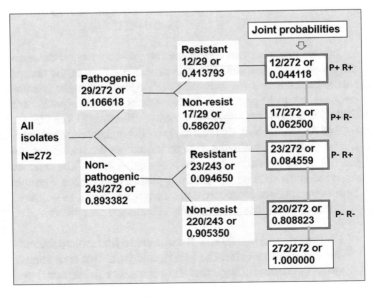

Fig. 2.18 *Staph.* isolations displayed as a tree

type of solution is being sought. In all cases, the total of the joint probabilities should add to 1.0000.

Note that the first intersection, which divides *pathogenic* from *nonpathogenic* is <u>not</u> conditional and corresponds to the right marginal totals in the table (Fig. 2.17). All other probabilities downstream from the first intersection are *conditional* upon the events that occurred previously.

Recall that **P(AandB)** is actually the product of **(A)** and **(B|A)** and is defined as a "joint probability." All joint probabilities are shown in boxes on the right side of the tree. (Joint probabilities also correspond to the values in the four central cells of a 2 × 2 table.)

To obtain the joint probability **P(AandB),** in this example, **P(pathogenic and resistant**), requires you to *multiply along the branch*, meaning that you are intuitively calculating **P(A) × P(B|A):**

$$\mathbf{P}\left(\textbf{pathogenicand resistant}\right) = \mathbf{P}\left(\textbf{path}\right) \times \mathbf{P}\left(\textbf{res|path}\right)$$
$$= 0.106618 \times 0.413793$$
$$= 0.044118$$

In case you were wondering, we cannot simply enter the overall known probability of *resistant* vs. *not resistant* at the second intersections unless we know for certain that resistance is *independent* of pathogenicity. In this example, **P(Res|path)** and **P(Res|nonpath)** are obviously quite different (0.413793 vs. 0.094650), confirming that resistance is dependent upon pathogenicity; that is, the pathogenic status influences the resistance status. This also confirms the cross-multiplying test of dependence from the same data on the previous page. For a further example comparing independence and dependence, see Fig. 2.19 below.

Other probabilities can be easily calculated from this tree.

- **P(res|nonpath)** = 0.09465. Remember to first isolate the conditional quantity (after the | line), and from that new sample space (as denominator), find the probability (numerator) you are looking for. In this example, start with **nonpath**, and from that branch, find the probability of resistance.

- **P(AandB)** is the (joint) probability of a single isolate being <u>both</u> pathogenic <u>and</u> resistant: = 0.044118
- **P(non-pathogenic):** 0.893382
- **P(resistant)** = for this we need to add both joint probabilities that have **R+** outcomes (the top and the third from the top). So 0.044118 + 0.084559 = 0.128677
- **P(nonresistant | nonpathogenic):** Start on the nonpath branch, then find P(nonres) = 0.905350. Note that finding a value when starting with a conditional probability on the left is simple.
- **P(pathogenic|resistant):** This is a little more complicated. The "condition" (resistant) is on the right side of the tree. First find and add ALL joint probabilities marked **R+**. They become the denominator. The numerator is the <u>one</u> joint probability that is both resistant <u>and</u> pathogenic: (0.044118) / (0.044118 + 0.084559) = 0.34286
- **Are A and B independent?** No. Resistance is *dependent* upon pathogenicity (= *"not independent"*). In other words, the likelihood of resistance changes if the isolation was pathogenic. You can see this by comparing P(Res|Path) and P(Res|Nonpath).

To further illustrate the meaning of independence, consider Fig. 2.19. The left (a) side of the diagram shows event A (history of a medical condition) and event B (reaction to a chemical) are independent.

Regardless of whether A has occurred or not, a reaction to B can be expected at about 10% of the time. But in Fig. 2.19b, reaction to B is much greater (60%) when A is positive but only 5% if A is negative. Therefore, A influences the probability of B. Event B **is** dependent upon A in the right-side diagram. *Hence, they are* NOT independent.

(a)   (b)

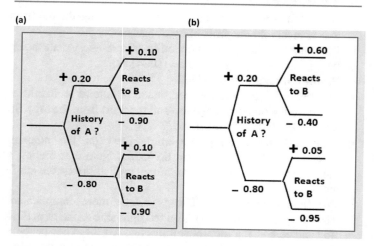

Fig. 2.19 (a) B independent of A. (b) B dependent upon A

2.8   Using Frequencies or Probabilities

The tree diagram in Fig. 2.18 shows both counts/frequencies as well as probabilities at each intersection and the end of each branch. Although probabilities are generally preferred, either can be useful, depending upon the type of problem, the data available, and the objectives.

2.9   Order of Entering Variables Into the Tree

As a guide, always read the question carefully to learn if the events actually occur in a sequence, first one then another. Does one decision, event, or occurrence logically happen first? For instance, if a gas leak must happen first for it to cause an explosion and then a fire, this is the order that they should be entered into the tree.

In the *Staphylococcus* example (Fig. 2.18), or in the hiring of three people (see Fig. 2.20), the outcome will be the same regardless of the order for constructing the tree. However, in these two examples, although the sequence doesn't matter, we still enter them one at a time.

In Fig. 2.20, three people are being hired, and each is tested for a genetically linked sensitivity to a group of chemicals being used in the workplace. We need to know the probability of this sensitivity being present in none, one, two, or all three individuals.

Each candidate is considered in turn, NOT as three branches from a single point. Remember that the total of the probabilities branching from a single point must add to unity (1.00). This example also demonstrates the advantage in labelling the joint probabilities for easy identification (e.g., SNS, NNN, etc.). This is also an excellent example of each person being independent from the other: the P(sensitivity) is the same whether previous individuals were sensitive or not.

Question (a) asks for P(*exactly one*) of the three candidates with sensitivity. Looking at the labels, we see that this is the total of 4th, 6th, and 7th joint probabilities from the top, each of which shows one "S." (b) asks for exactly two, requiring the 2nd, 3rd,

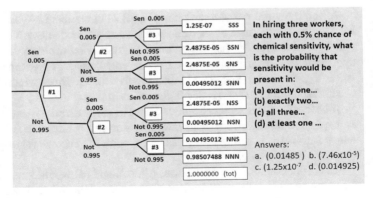

Fig. 2.20 Sensitivity is independent of the order of selection

and 5th joint probabilities to be added together. In (d), "at least one" is *all* the joint probabilities added except the lowest one, OR (the faster method using complements) the total (1.000) minus "none" (0.98507488) = 0.014925.

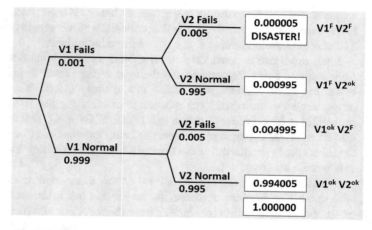

Fig. 2.21 Failing valves in sequence

A liquid natural gas storage plant (Fig. 2.21) transfers LNG to road and rail containers using a series of two motorized valves V_1 and V_2. Valve type V_1 has been found to have a failure rate of approximately once in a thousand times when used. A backup valve, V_2, is an older design and fails roughly 1 in 200 times when used. The failure rates are independent of each other.

If both valves fail, a catastrophic release of pressurized gas and an explosion would occur. This example presents a situation in which one valve is brought into play first, with the second valve playing a backup role.

The order counts. While questions (a), (b), and (d) would have the same answer, (c) is meaningless if the order of the valves was switched.

(a) Find the probability of *at least one valve failing* during a single transfer of LNG. In a simple data array such as this, the question is asking you to add the probability of BOTH failing, + the prob. of just the first one failing + the prob. of just the second one failing.

Add these three probabilities:

$$0.000005 + 0.000995 + 0.004995 = 0.005995.$$

A faster and simpler way would be to recognize that "at least one" is really ALL probabilities except the "NONE" option. So: $1 - 0.994005 = 0.005995$

(b) Find P(both valves failing)? Solution (visually): 5×10^{-6} or 0.000005

or using the formula: $P(A \text{ and } B) = P(A) \times P(B|A)$:

$$P\left(V_1^F \text{ and } V_2^F\right) = P\left(V_1^F\right) \times P\left(V_2^F \mid V_1^F\right) = \left(1 \times 10^{-3}\right) \times \left(5 \times 10^{-3}\right)$$
$$= 5 \times 10^{-6} \text{ or } 0.000005$$

(c) Given that V_1 fails, what is the probability that V_2 will also fail during the same transfer event? Note that here, the failure rate of V_2 is independent of V_1 failing, meaning $P(V_2^F|V_1^F) = P(V_2^F)$. Simply start not at the left side but move along the "V_1-fail" branch. Solution: $P(V_2^F|V_1^F) = 0.005$ (this was actually given).

(d) The facility transfers LNG 30 times/week for 52 weeks/yr. What is the probability of at least one release of LNG (= both valves failing) during the projected 10-year life of the facility [$30 \times 52 \times 10 = 15{,}600$ events]?
Solution: $1 - \{P(V_1^F \text{ and } V_2^F)^C\}^{15,600} = 0.075$ or 7.5%. For the method used here, please see Sect. 2.12.

2.10 Probabilities Expressed in Disease Screening

2.10.1 Disease Screening Using Bayes' Theorem

Individual outcome probabilities can be strengthened by including *prior information about the population* into the calculation. This underlying theory was established by Thomas Bayes in the early 1700s.

We borrow from Bayes' principle to analyze the probabilities following testing for a condition or disease. For this, we begin with a known population incidence as the first decision intersection (left side) and the results of individual test results (positive/negative) as the second intersection (right side). Joint probabilities can then be calculated.

Figure 2.22 depicts the incidence among the population of a disease condition as 2 percent (0.02).

The test given to individuals is imperfect, and in this instance, the test gives a positive reaction in only 97 percent of the people

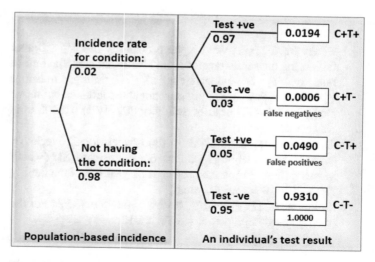

Fig. 2.22 Use of Bayes' theorem for interpreting disease screening

who <u>have</u> the condition. For the remaining 3% of the positive people, the test shows, falsely, a negative result: the *"false negatives."* The test also gives a *"false-positive"* reaction in 5% of the people who do <u>not</u> have the condition.

The tree is drawn such that the top joint probability represents the "true" positives and the lowest the "true" negatives. The second from the top are the "false negatives" (really positive but showing negative – the dreaded combination for infectious disease control!), and the third from the top are the "false positives" (really negative but showing positive). The following questions are commonly asked by patients, while physicians often find that they are often not well prepared to convey this type of information accurately.

(a) Given that an individual has just received a positive reaction from this test, what is the probability that the individual has the condition? Note that the wording is presented in a slightly different order, but the meaning is clear: *"What's the probability that the individual has the condition, given that they have a positive test?"* We write this as: **P(C+|T+)**?

Begin with the "test-positives" (T+) as the (reduced) sample space or denominator. For this, you add the top and third joint probabilities. They are (**C+ and T+**) and (**C– and T+**). (The second and fourth are **T–** and not relevant). Add these two joint probabilities together to obtain 0.0684. This total with **T+** results is our reduced sample space and our denominator. The numerator is determined from the left side of **P(C+|T+)** and is clearly the top joint probability **P(C+ and T+)**. The calculation therefore becomes:

$$P\left(C+\mid T+\right) = \left(0.0194\right)/\left(0.0194 + 0.0490\right) = 0.2836.$$

The interpretation is that a person who has been given a positive screening test result has only around 28 percent chance of actually having this condition. This screening test has a fairly high false-positive rate. Subsequent detailed testing will provide more accurate results, but these mass screening tests provide a useful first step.

(b) *Find the probability that a person is really positive even though they have been tested as negative.* This is a conditional probability. It is asking you to find **P(C+|T−).** The denominator is the total probability of having a negative test (add 0.9310 and 0.0006). The numerator is the specific subgroup we are looking for (C+) among the test-negative group:

$$P\left(C+ \mid T-\right) = \left(0.0006\right) / \left(0.0006 + 0.9310\right) = 0.00064.$$

A person who has a negative test result can rely on it; there are only slightly more than 6 chances in 10,000 that her test is false and that she actually has the condition. Consumers of health care can generally rely more on the accuracy of a negative screening test result than on the accuracy of a positive test result.

The tree in Fig. 2.22 also provides answers to other important questions. (Be very careful about the exact wording):

(c) *Given a person has the condition, what's the probability of a positive test?*

..........P(T+|C+) = [0.9700]

(d) *Given a positive test result, what's the probability they have the condition?*

(Calculation is shown above) ....P(C+|T+) = [0.2836]

(e) *Given a person does NOT have the condition, find the probability of a positive test?*

(This was given initially)..........P(T+|C−) = [0.0500]

(f) *Given a negative test result, what's the probability a person does not have the condition?*

................P(C−|T−) = [0.99936]

(g) In a population of 1 million, with this 2% (0.02) incidence, how many people would actually have the condition?   ....1000,000 × 0.02 = [20,000]

(h) In a population of 1 million, how many people would test positive:

.....$1 \times 10^6 \times (0.0194 + 0.0490) = 1,000,000 \times 0.0684 = [68,400]$

(i) In a population of 1 million, how many would test negative but would really have the condition?

..........$1 \times 10^6 \times 0.0006 = [600]$

2.10.2 Disease Screening Using Sensitivity and Specificity

Anyone in the medical or epidemiology fields is familiar with explaining false negatives and false positives in terms of *specificity* and *sensitivity*. Figure 2.23 compares a common (but less accurate) mass screening test, with the limited-use, "gold standard" test by which other tests are compared.

The *sensitivity* of the mass screening test is calculated as the proportion of the true positives that were correctly detected as positive by the test. In this case, 840/1000 or 0.84. The *specificity* of the screening test is calculated as the proportion of the true negatives that were correctly detected as negative by test. In the example, this is 900/1000 or 0.90. In reality, we test the target population without knowing at that time the accuracy of the individual test.

This chart also allows the estimating of the proportion of the screened positives that can be assumed to be truly positive (the "positive predictive value" or PPV), as well as estimating the proportion of the screened negatives that can be assumed to be truly negative (the "negative predictive value" or NPV).

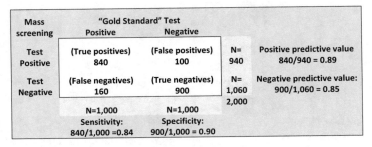

Fig. 2.23 False positives and false negatives in specificity and sensitivity

2.11 When a Variable's Independence Is Missing or Uncertain

Consider the following situation, we are told that 1 in every 150 children are allergic to nuts, and 1 in every 250 children are allergic to tartrazine food color. We do not know if these allergies are independent. We DO know that 1 in 350 children have BOTH allergies.

Clearly, the first probability (Fig. 2.24) is 1/150, but we cannot assume that the probability at the second intersection will be 1/250 because we do not know if the second variable is independent of the first. The probability of tartrazine allergy overall is 1/250, but those with or without nut allergy may have different probability of tartrazine allergy.

We are told the probability of having both allergies, P(N+ and T+), is 1/350 or 0.002857. This is the top joint probability, and it also allows us to find P(T+|N+): 0.002857/0.006667 = 0.428529.

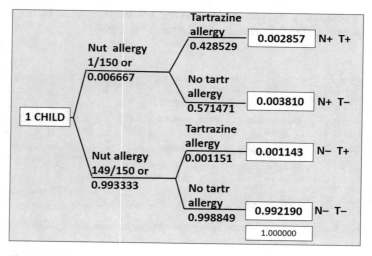

Fig. 2.24 The nonindependent (= dependent) second variable

The complement of this $(1 - 0.428529)$ is 0.571471, which is $P(T-|N+)$.

The top and third joint probabilities each show **T+** and when added together represent total tartrazine allergy of 1/250 or 0.00400. We already have the top one (1/350), so the other must be 0.001143, and from this, the remainder of the conditional and joint probabilities can be found. (Remember if $A \times B = C$, then $C/A = B$)

From Fig. 2.24, find the probability of selecting one child, at random, who is:

- *Allergic ONLY to nuts?* (N+T−) **0.003810**
- *Allergic to nuts OR tartrazine (or both)?* ..... $0.002857 + 0.003810 + 0.001143 = $ **0.007810** (*or we could have used* $1.000000 - 0.992190 = $ **0.007810**)
- *Allergic to NEITHER nuts NOR tartrazine?* **0.992190**
- *What is the probability* that a class of 50 children will contain at last one child who is allergic to either of the two foods? ..... **0.3243** (or 32.4%). Please see Sect. 2.12 for the method used here.
- *Are the two allergies independent?* No. $P(T+|N+)$ is different to $P(T+|N-)$ so having nut allergy affected (increased in this case) the probability of tartrazine allergy; hence, these variables are dependent.

2.12   Calculation of "At Least One" Outcome (with Multiple Iterations)

Several previous examples have referred to this section to explain a somewhat complex calculation that runs contrary to our initial intuition. For example, if the probability of a failure is one in a hundred, and you ask someone what would be the probability of a failure after 25 outcomes, they may respond $25 \times 0.01 = 0.25$ or one in four. (This is close, but not exact). What if there are 100 outcomes? Is that "certainty" or $100 \times 0.01 = 1.00$? Of course not, because there is always a chance it will not happen, even after 150 attempts or 1000. And after 150 attempts, is the probability of it happening 1.50? This is obviously wrong, as probabilities exist only between 0.00 and 1.00!

Multiplying the probability for one event by the number of events is clearly false. Depending upon the narrative and values used, this approach can sometimes result in an approximation of the correct answer, but the error increases with the number of iterations of the event.

Taking again the data from Fig. 2.21, the facility transfers liquified natural gas (LNG) 30 times each week for 52 weeks a year. We ask: *"What is the probability of at least one disastrous release of LNG (= both valves failing) during the projected 10-year life of the facility?"*

From Fig. 2.25a, we see that only the top probability (0.000005) is the "disaster" probability for one event. Step 1 is to find the complement of that. Thus, 0.999995 is the probability that NO disaster in <u>one</u> event (equal to the other three joint probabilities added together).

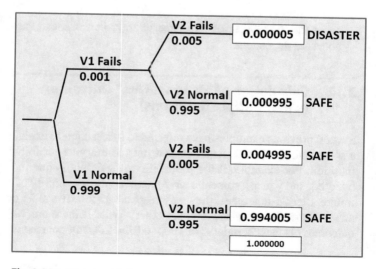

Fig. 2.25a Disaster risk for one event

Figure 2.25b shows the tree redrawn this time with <u>three</u> events and the probability just of "disaster" or "no disaster" after each event. Observe that with only three events, eight joint probabilities have appeared. All carry some likelihood of disaster *except for the bottom one.*

P(no disaster after 3 events) = 0.999985

Important

Although we are ultimately seeking the probability that *"at least one"* disaster will occur in N events, the initial approach is to determine the probability *of NO disaster* in N events. Figure 2.25b demonstrates the **P(no disaster in 3 events)** by raising 0.999995 to the power of three, resulting in 0.999985. The complement of this figure (= the sum of all the other joint probabilities) is "at least one disaster."

This is the solution for three events, but the question asked for 10 years' worth of events. If we project this calculation to 10-year life of the facility (during which $30 \times 52 \times 10 = 15,600$ transfer

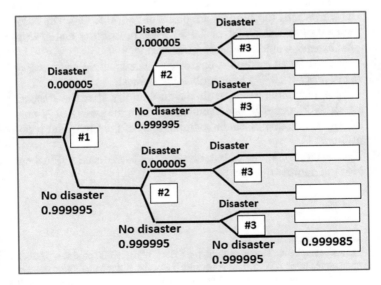

Fig. 2.25b Disaster risk for three events

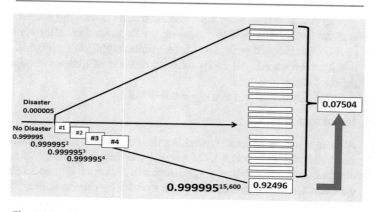

Fig. 2.25c Disaster risk after 15,600 events

events will occur), with 0.000005 chance of disaster for each, the
P(no-disaster) in 15,600 events is 0.999995 raised to the power
of 15,600 (Fig. 2.25c).

This diagram illustrates the risk of NO disaster after 15,600
transfer events, reaching a joint probability of **0.92496.** The complement of this (**0.07504**, or **7.5%**) is the probability that *at least
one disaster* would take place in ten years[2].

This shortcut method is equivalent to calculating *every one of
the other* ($1 \times 10^{4,596}$) joint probability values!

The technique starts with the probability that the disaster
would NOT occur in one event, raises it to the power of N possibilities, and then takes the complement of the final joint probability.

In brief format, the calculation is: $1 - \{1 - P(disaster)\}^N$ where
N is the number of events.

---

[2]In this example, incorrectly multiplying 5.0×10^{-6} by 15,600 produces 0.078, a
value quite close to 0.075. This is because the probability of a fault is extremely
small. This approximation could be used in such a case, but using the correct
method is preferred for avoiding errors that can be embarrassingly large.

2.13   Probabilities: A Summary

1. A probability is a number between 0 and 1.
2. A sample space is a list of all possible outcomes of an event.
3. Each outcome has its own measured probability of occurring.
4. All probabilities at each intersection add to 1.00.
5. Addition rule for mutually exclusive events: **P(A or B) = P(A) + P(B)**
6. Addition of any A and B events: **P(A or B) = P(A) + P(B) – P(A and B)**
7. **A^c** = complementary event: **P(A^c) = 1 – P(A)**
8. Event **A** followed by **B** is written **P(A and B)** and obtained by multiplying (A) by (B given that A has taken place). Thus, joint events **P(A and B) = P(A) × P(B|A)**
   Or usefully rearranging: **P(B|A) = P(A and B)/P(A)**
9. If A and B are independent events, then: **P(A and B) = P(A) × P(B)**
10. Branches from the first intersection are non-conditional probabilities, for example, P(A); subsequent branches are all conditional probabilities, for example, P(B|A).
11. In trees: multiply along branches; add or subtract between branches
12. To find "at least one" in large-N events, use: $1 – \{1– P(event)\}^{N\text{-possibilities}}$

2.14   Case Study #2: Challenger Shuttle Disaster

Case Study #2: Challenger Shuttle Disaster
The historical failure rates of early solid-fuel rockets from the time of Goddard had been about **1 in 57**, but modern authorities were more optimistic:

- **1983: An independent risk assessment** claimed a solid-fuel booster will fail **1 in 70**.
- **1984: The US Air Force** reported chance of a single booster failing to be **1 in 210**.

- **1985: NASA** announced risk of shuttle failure **1 in 100,000** launches.

Weeks before the space shuttle Challenger exploded a few seconds after launch on January 28, 1986, a panel of professional risk analysts told a congressional subcommittee that they had calculated that the shuttle's solid rocket boosters had a much higher chance of failing than NASA engineers claimed. After reviewing failure rates of 1903 solid-rocket launches, they estimated the likelihood of a booster failure was closer to **1 in 1000**. NASA mistrusted these calculations, preferring to use their own engineering judgment to reach a "design objective" of one failure in 100,000 uses (Los Angeles Times 1986).

Challenger Challenge: Using 1 in 57 (the early, worst-case failure rate for single booster), how many launches would it take to reach a 50–50 probability of total loss for a single launch with two boosters? (Remember one or both failing would cause total loss).

Total loss would happen if either one or both solid-fuel boosters failed. So the probability of ANY failure is 3.06×10^{-4} plus 2×0.01719.

That total is 0.03469 for a single launch. The probability of no failure for a single launch is $1 - 0.03469$, or 0.9653. Raising this to N launches will enable us to predict disaster.

After only 20 launches (Fig. 2.26b), using the historical evidence for failure of solid-fuel rockets, failure was predicted to be 50% likely ($0.9653^{20} = 0.4935$, closest to 0.5000).

On January 28, 1986, a few seconds after the 25th launch, a booster failed with the loss of all lives.

Conclusion: The 1 in 57 historical model predicted a 59% chance of total loss after 25 launches.

Of course, a single incident such as this is neither valid nor reliable, but the observation remains that coincidence or not, this prediction was surprisingly good, and a very long way from the "1 failure in 100,000 launches" predicted by NASA just the previous year!

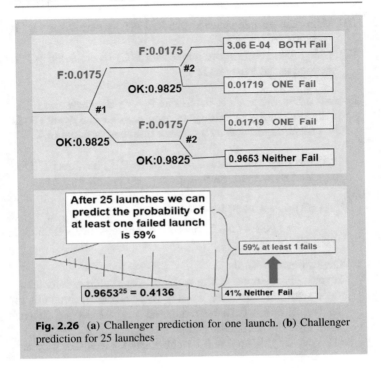

Fig. 2.26 (a) Challenger prediction for one launch. (b) Challenger prediction for 25 launches

Practice Exercises: Chapter 2

1. Heads or Tails

1	2	3	4	5	6	7	8	9	10	11	12	13	14	15	16
H	H	H	H	T	H	H	H	T	T	T	H	T	T	T	T
H	H	H	T	H	H	T	T	H	H	T	T	H	T	T	T
H	H	T	H	H	T	H	T	H	T	H	T	T	H	T	T
H	T	H	H	H	T	T	H	T	H	H	T	T	T	H	T

[A] Using the sample space above, find the probability of event "A" occurring, where "A" = "at least two tails." This can be written as P(A) ≥ 2 tails.

[B] Using the sample space above find P(B) where event "B" is
">2Tails"

[C] Using the sample space above find P(C) where event "C" is
defined as "at least one H"

2. **Toss a Die.** We wish to determine P(A or B) where event (A)
is the probability of rolling an even number and event (B) is
the probability of rolling a number less than or equal to 4.

P(A) has three outcomes: 2, 4, 6, while P(B) has four: 1, 2, 3,
4. (A and B) = 2, 4, or 2/6

$$P(A \, or \, B) = P(A) + P(B) - P(A \, and \, B)$$
$$= 3/6 + 4/6 - 2/6 = 5/6 \text{ or approximately } 0.8333$$

3. **Contaminated Wells**
Seventy-five percent of the wells in a township are contaminated with fecal bacteria. Twenty-five percent of the wells in
the township are both contaminated and chlorinated (hence
safe). Find the probabilities that:
   (a) Wells in this area will be chlorinated, given that they are
       contaminated.
   (b) A well drawn at random will be contaminated but unchlo-
       rinated.
   (c) A well drawn at random will be uncontaminated. (Assume
       that a well is still considered contaminated, even though it
       may have been chlorinated).

4. **Chemical Allergy**
About 8% of people in a very large population have an allergic
reaction to materials used in the production process at a large
chemical plant. If two people selected at random from the population are hired by the company tomorrow, find the probability that (a) neither is allergic to the materials, (b) both are
allergic to the materials, and (c) exactly one is allergic to the
materials.

5. **Safety Violations Analysis**

   Company records show that 80 employees have been reprimanded for safety violations more than once, and 300 have
   never been reprimanded. The company has 400 employees.
   Those who have zero or one reprimand on their records are
   evenly split between men and women. The people who have
   more than one reprimand are 75% men. Determine the probability (use a tree) of randomly selecting (a) a male employee
   with one reprimand, (b) a male, and (c) any person with no
   reprimand given that the person is a male.

6. **Performance Projections**

   A human resources manager knows from experience that 80%
   of applicants for a job will be able to perform satisfactorily. Of
   those who do satisfactory work on the job, 90% pass an abilities test. Of those applicants who do not perform satisfactorily
   on the job, 60% do not pass the abilities test. If an applicant
   selected at random passes the test, find the probability that (a)
   the applicant will perform satisfactorily, and (b) the applicant
   will not perform satisfactorily.

7. **Health Status of Tourists**

Gender	Ill	Not ill	Total
Male	47	123	170
Female	43	122	165
Total	90	245	335

   The accompanying contingency table classifies the health status of 335 visitors to a Central American country. What is the
   probability that:

   (a) A randomly selected person is a woman, given that they
       are ill.
   (b) Any individual will remain well given that they are female.
   (c) A randomly selected individual is a man.
   (d) Is illness independent of gender?

8. **Pacemakers**

A firm is hiring three employees to work with microwave transmitters. One in every 500 people of this age group wears a heart pacemaker and thus would be vulnerable to microwave interference. What is the possibility that [a] *none* of the three will wear a pacemaker, [b] *at least one* employee might wear a pacemaker, and [c] *exactly one* of the three will wear a pacemaker. (Keep the precision to SIX decimal places).

9. **Pipeline Valves**

An oil pipeline has three safety shutoff valves in case the line springs a leak. These valves are designed to operate independently of one another. There is only a 7% chance that valve #1 will fail at any one time, a 10% failure chance for valve #2, and a 5% failure chance for valve #3. Find the following probabilities (a) that *all three valves operate* correctly, (b) that *all three valves fail* when needed, (c) that *exactly one* valve operates correctly, and (d) that *at least one* valve operates correctly.

10. **Safety Training**

A chemical company has 800 employees. Twenty-two percent of the employees have had training, and exactly half of these people are in nonmanagement positions. Exactly one-third of the non-trained people are in management positions. (a) How many managers does the company have? (b) What is the conditional probability of being trained, given that the person is a manager? (c) What is the conditional probability of being a manager, given that the person has had training? (d) Find the probability of randomly selecting an untrained person among the managers. (e) Find the probability of finding a person at random who is both trained and a manager.

11. **Reliability of Defective Sensors**

Quality testing of a large batch of heat sensors shows that 15 percent are defective and will fail one time in 4 when needed. Normally, these sensors fail only 1 percent of the time when needed (the remainder (85%) of the valves are of this type). (a) From this batch if a sensor fails, what is the probability it is a faulty type? (b) Given there are 20,000 sensors in the batch, and on average each is required to detect excessive heat

10 times in a 5-year life expectancy, how many **additional failures** to detect excessive heat will result because of the defective sensors over that 5-year period?

12. **Oil Spill Projections**

From studies of oil spills at similar oil loading facilities, assessors feel that there is a 0.6 probability of an oil spill into the sea and a 0.8 probability of an oil spill on land during the first year. What is the probability of (a) both a sea and a land spill during the first year? (b) Of only one spill? (c) Of neither?

13. **Decision: Dialysis**

The kidneys of a patient with end-stage kidney disease will not support life. If the patient is to survive, the available choices are a kidney transplant or the use of a dialysis machine several times per week. The patient is faced with the choice. Her physician gives her the following information: About 68% of dialysis patients survive for 5 years. Of transplant patients, about 48% survive with the transplanted kidney for 5 years, 43% must still undergo regular dialysis because the transplanted kidney fails, and the remaining 9% do not survive the transplant. Of those transplant patients who return to dialysis, about 42% survive 5 years. Which treatment should the patient choose to give herself the best chance of living for 5 years?

14. **Screening for Prostate Cancer**

Cancer of the prostate is the number two killer of North American men. About 1 in 10 men will develop the cancer in his lifetime. A blood test has been developed for early detection of this cancer, and the Cancer Institute plans an evaluation study of the screening technique. The annual incidence rate is about 1/500 patients, or the probability that an unsuspecting man has this type of cancer, P(cancer) = 0.002. Unfortunately, the screening technique is not 100% effective and results vary between patients. The probability of a positive test given that there is no cancer has been estimated to be about 0.16. The probability of a negative test, given that the individual has cancer, has been estimated to be about 0.20.

Find the probabilities (a) of having this cancer given a positive test result, (b) of not having cancer given a positive test result, (c) of having the cancer given a negative test result, (d) of a positive test result, and (e) of testing positive given that a person has the cancer.

15. **Screening for Cervical Cancer**

Cervical cancer accounts for around 4% of all cancers diagnosed in women and is responsible for 3.5% of all cancer deaths. The recommendation is that women over the age of 20 be screened at least every 3 years by cervical cytology. However, the effectiveness of cervical cytology has never been tested in an experimental study. The annual incidence rate for this cancer is about 1.1 in 1000 patients, or a patient randomly selected for screening has a 0.0011 probability of having this cancer.

The screening test is not perfect; the probability of a positive test given that there is no cancer has been estimated to be about 0.006. Cytology can also result in a negative test or no recommendation for further diagnosis, given that there is actually cancer present, and this has been estimated at a probability of 0.3 Find:

[a] The probability of having this cancer, given a positive result

[b] The probability of not having the cancer, given a negative result

[c] The probability of not having the cancer, given a positive result

16. **Nuclear Reactor Safety**

A nuclear power reactor has two "fast shutdown" (FSD) systems. The first system is unavailable due to component failure or maintenance for an average of 20 days per year, and the second system for 25 days per year. The situation requiring a fast shutdown has an annual probability of 0.1, and the risk of a "serious incident" if the shutdown process is not available is 50%. Calculate the following:

[a] The annual prob of the need for an FSD and at least one system responding correctly.

[b] The annual prob of the need for an FSD and neither system being available.

[c] Given the need to shut down, what is the prob that at least one FSD system will be working?

[d] The annual probability of a "serious incident" occurring.

[e] The expected life of this reactor is 20 years. What is the prob of a "serious incident" occurring in its lifetime?

17. **Cholera in the Andes**

During an outbreak of cholera in a mountainous region in Peru, 15% of the 12,000 people in the region contracted the infection. Due to distances and late notification, only 80% of the cases received treatment. The case-fatality (CF) rate for treated cases is approximately 1%, whereas non-treated cases generally experience a CF of 50%.

[A] What was the cholera-specific mortality rate? (Total deaths)/(Population)

[B] What was the chance of dying from cholera, given that you were a case?

[C] What was the chance of having had treatment, given that you survived an infection?

[D] Estimate the number of survivors who owe their lives to treatment.

[E] Estimate the number of deaths that may have been prevented, had treatment been available for all.

18. **Dust Explosion and Fire**

The annual risk of an explosion due to dust in a mill is one in a hundred. Previous experience has shown that there is an 80% chance that the explosion will start a fire. A sprinkler system has been installed, which is likely to function 99 times out of 100. Whether the sprinkler system works or not, the fire alarm will be activated with only 1 in 1000 chances of failure. Show these events by means of a "tree." Calculate the following (showing results as probabilities ($0 < P < 1$)):

(a) Annual risk of not having an explosion

(b) Annual risk of explosion + subsequent fire

(c) Annual risk of explosion + fire + sprinklers working + fire alarm activated

(d) Annual risk of explosion + fire + EITHER sprinklers OR alarm working (or BOTH)

(e) Annual risk of explosion and fire, but failure of BOTH sprinklers and alarm

(f) Given that there has been explosion and fire, what is the probability of BOTH the sprinkler system AND the fire alarm failing?

(g) Given that there has been an explosion and fire, what is the probability that at least ONE (sprinkler or alarm) will work?

(h) Annual risk of explosion without a subsequent fire?

19. **Blood Screening for AIDS**

Prior to March 1985, several hundred people were victimized by AIDS when they received transfusions of blood that contained HIV. At the time about 4 in 10,000 units of blood were infected with HIV. Routine testing of blood donations for antibodies to HIV began in March 1985, and administrators began stating that the public need have no worry about getting AIDS from the blood supply. Unfortunately, tests for HIV are not error-free. The donor may be infected, but with an immune system not having recognized the virus. It is estimated that 1 in 100,000 current blood donations are test negative, but virus positive. About 1 in 1000 units test positive when the blood does not have the virus present. Using an appropriate probability tree, [a] find P(test neg | blood pos), [b] find P(test neg | blood neg), [c] find P(blood pos | test neg), [d] find P(blood pos | test pos), [e] if 800,000 units of blood are collected each year, how many accidental cases would be expected in the 10 years between 1985 and 1995?

20. **Emergency Power in a Hospital**

Westtown hospital has two generators for emergency power. Historical records show that there is a 3% chance that the first generator will not be available when needed (due to mechanical fault or maintenance) during a total power failure, and an 8% chance of the second one will not be functioning. A surgical operation is scheduled for about 15% of the time, and in 70% of these, a power failure would result in a fatality (e.g., patient is on a heart-lung machine).

[a] Find the probability that in a total power failure, neither generator will operate. (Note: If one generator fails, it will not influence the failure of the other one).

[b] The hospital has an expected life of 25 years, and there are, on average, 20 local power failures per year. What would be the expected risk of such a fatality over the life of the hospital?

[c] By how much would that risk be reduced by the addition of a <u>third</u> generator (with reliability of 98%)?

[d] If the expected settlement in a lawsuit is $8 million for such a fatality, how much would the average pay-out risk (liability) over 25 years drop with the purchase of the third generator?

21. **Aeroengine Malfunction**

A certain type of piston-driven aeroengine has been found to malfunction in normal flying conditions on average once in 8000 hours.

[A] What is the probability that a single engine will fail during a 3-hour flight?

[B] A twin-engine aircraft is designed to fly even with one engine dead. What is the probability that a twin-engine aircraft will experience exactly one engine failure during a flight of 3 hours?

[C] A forced landing is imminent if both engines fail. What is the probability that the aircraft will experience shutdown of both engines in a 3-hour flight? (Assume the failures are independent of each other).

[D] An aircraft charter company in the Arctic begins operation on July 1, 1996, and makes an average of 10 flights a week with a twin-engine aircraft, and each flight an average of 3 hours. On what date (approximately) would the company reach a 0.5 probability of a failure of one engine?

22. **Long Distance Trucks Maintenance**

The province licenses 60,000 long-distance transport trucks. Province-wide checking reveals that only 68% of all these trucks are correctly maintained to the ministry's standard. During the year, 1800 trucks have been involved in accidents. Accident investigators have determined that 56% of trucks involved in an accident have not been correctly maintained.

[a] Of trucks involved in accidents, what percentage had been correctly maintained?

[b] How much more likely is an improperly maintained truck to be involved in an accident than one that is correctly maintained?

[c] What percentage of trucks overall are involved in accidents?

[d] What percentage of incorrectly maintained trucks are involved in accidents?

[e] How many accidents would be prevented by increasing the maintenance rate from 68% to 75%?

23. **Chemical Sensitivity**

Extreme chemical sensitivity is a life-threatening situation that can be fatal in a few minutes. When the normal Y5 gene appears in a mutated form (we can call the condition Yx), there is no apparent effect on the individual's sensitivity to environmental chemicals, but if the individual ALSO has a defective E enzyme (we can call this condition Ex), then the individual has a life-threatening sensitivity to certain plasticizers in common use in industry and the environment. (Separately, these factors do not increase a person's risk at all. BOTH factors must be present). One person in every 340 has the mutant gene, and 1.5% of the population has the defective enzyme. We do not yet know if having the defective enzyme is independent from having the mutated gene. A recent survey shows that 53 in 120,000 persons are carrying both the mutated gene AND also the defective enzyme. Construct a tree with these data.

(a) What proportion of the population is free of BOTH factors?

(b) What proportion of the population carries the mutated gene but NOT the defective enzyme

(c) Are the mutated gene and the defective enzyme independent or dependent factors?

(d) For those with the mutated gene, what is the risk of having the defective enzyme?

(e) For those with the defective enzyme, what is the risk of having this life-threatening chemical sensitivity?

(f) In this population of 120,000 persons, how many would be at risk of such a sensitivity?

Quest	Solution and comments
1	[a] 11/16 = 0.6875, [b] 5/16= 0.3125, [c] 15/16 = 0.9375
2	$P(A \text{ or } B) = P(A) + P(B) \text{——} P(A \text{ and } B)$ $= 3/6 + 4/6 \text{——} 2/6 = 5/6$ or approximately 0.8333
3	[a] P(Chl|contam): 0.25/0.75 = 0.333′, [b] P(contam and unchl): 0.75 − 0.25 = 0.5, [c] P(uncontam): 1 − P(contam) = 1 − 0.75 = 0.25
4	[a] 0.8464, [b] 6.4×10^{-3}, [c] 0.1472
5	[a] P(male and 1 rep): 10/400 = 0.025, [b] P(male): 220/400 = 0.550, [c] P(norep|male): 150/220 = 0.682
6	[a] P(sat|T+): 0.72(0.72 + 0.80) = 0.90, [b] P(unsat|T+): 0.08/(0.72 + 0.08) = 0.10
7	[a] 0.48, [b] 0.74, [c] 0.51, [d] OR = 1.08, close to 1.0: illness independent of gender
8	[a] 0.994012, [b] 0.005988, [c] 0.005976
9	[a] 0.79515, [b] 0.00035, [c] 0.01445, [d] 0.99965
10	[a] 296, [b] 0.2973, [c] 0.5000, [d] 0.7027, [e] 0.1100
11	[a] 0.8152, [b] 7200 additional failures
12	[a] 0.48, [b] 0.44, [c] 0.08
13	Dialysis: 5y survival: 68.00%, transpl: 5y survival overall: 66.06% (dialysis slight advantage)
14	[a] 0.00992, [b] 0.99008, [c] 0.00048, [d] 0.16128, [e] 0.80000
15	[a] 0.113848, [b] 0.999668, [c] 0.886152
16	[a] 0.099625, [b] 0.000375, [c] 0.99624695, [d] 0.00018765, [e] 0.00374637
17	[a] 1.17%, [b] 19.8%, [c] 88.8%, [d] 706 lives due to treatment, [e] additional lives 176
18	[a] 0.99, [b] 8×10^{-3}, [c] 7.912×10^{-3}, [d] 7.99992×10^{-3} (round off to 8.0×10^{-3}), [e] 8.0×10^{-8}, [f] 1.0×10^{-5}, [g] 0.99999, [h] 0.002
19	[a] 0.025, [b] 0.999, [c] 1.0014×10^{-5}, [d] 0.28066, [e] 80 (av. 8 per year)
20	[a] 2.4×10^{-3}, [b] 11.84%, [c] with 3 gen risk: 0.25%, [d] liability drops from \$947 K to \$20 K
21	[a] 3.7495×10^{-4}, [b] 7.497×10^{-4}, [c] 1.40625×10^{-7}, [d] .5 prob after 924 flights (April 1998)
22	[a] 44%, [b] 2.7 times, [c] 3.0%, [d] 5.25%, [e] 139 fewer accidents
23	[a] 0.982500, [b] 0.0024995, [c] Dependent, [d] 0.150166, [e] 0.029444, [f] 442

References

Wilson R, Crouch EAC. Risk assessment and comparisons: an introduction.
    Science. 1987;236:267–70.
Wines M. NASA estimate of rocket risk disputed. Los Angeles Times, March
    5, 1986.

Quantitative Risk Assessment

3

Abstract

In this chapter we switch to chronic risks that usually accumulate slowly over years or a lifetime, and manifest as either carcinogenic or noncarcinogenic outcomes. The exposures are ingested (from food, water), inhaled (air, vapors, gases), absorbed through the skin, or taken as pharmaceuticals. Workplace exposure is of course high on the list of potential exposure routes, but ordinary environmental exposures at home, at school, or recreation are also represented in the examples.

3.1 Scope of Quantitative Risk Assessment

3.1.1 Long-Term Exposures

The calculations in this chapter are based upon chronic intake of relatively small quantities of substances or other agents over long periods of time, often a lifetime. The risks that arise from these exposures are divided into *carcinogenic risk* (which includes carcinomas, sarcomas, lymphomas, and leukemias) and *noncarcinogenic risks* (which can include reproductive and other toxicological outcomes or disorders.

© The Author(s), under exclusive license to Springer Nature
Switzerland AG 2023
E. Liberda, T. Sly, *Assessment and Communication of Risk*,
https://doi.org/10.1007/978-3-031-28905-7_3

3.1.2   The Maximally Exposed Individual (MEI)

We sincerely hope no one actually possesses all the characteristics and behaviors of the MEI. This hypothetical person is assumed to live a life in which the highest level of exposure is maintained throughout the study period or lifetime (taken as 70 years), essentially the "worst-case scenario" embodied in a single person. The rationale for basing corrective measures and precautions on an "extreme" exposure, rather than a "most likely" exposure, is that by addressing such an extreme set of conditions, the overwhelming majority of the population would be positioned lower on the risk scale. Measures, remedies, and precautions in place for the MEI would therefore ensure as far as possible that the average person would be less at risk and better protected. The prospect of anyone in the population having a lifetime exposure similar to or greater than the MEI is vanishingly small. Keenan discusses the pros and cons further in Sect. 3.6.[1]

3.1.3   The Four-Step Assessment Model

The measurement and estimation of carcinogenic and noncarcinogenic risks are commonly performed as the probability of the adverse effect (usually specified as mortality or morbidity) occurring within the lifetime of the individual. We will use a modified four-step method originally established by the US National Academy of Sciences of the National Research Council in what is commonly referred to as the "Red Book" (NRC 1983). It has been adopted by many countries as a standard approach to risk assessment. The four steps are (1) hazard identification: the identification and quantification of the substance(s) present at the point of the spill, leak, accident, release, or other circumstance; (2) dose-

---

[1]As an illustration, the probability tree analyses in Chap. 2 emphasized the *multiplication along the branch* of all variables. The analogy here is that as each separate "worst-case" factor which contributes to the overall exposure of the MEI is multiplied together, the joint probability will carry a very large overestimation error.

response assessment: the adverse effects that can be predicted through exposure to these substances, given the characteristics (age, activities, work, etc.) of the people exposed; (3) exposure assessment: examination of the pathways via which the original or transformed substances could directly or indirectly reach the human "receptors" and in particular the medium through which the receptors could be exposed (food, water, air, skin contact, workplace exposure, pharmaceuticals); and (4) risk characterization: the characterization and interpretation of the results of the analysis in terms of projected lifetime risk of carcinogenic or non-carcinogenic events.

In this book, the (US) NAS/NRC model has been adjusted (Fig. 3.1) to position *risk characterization* within both risk assessment and risk management. *Risk communication* is also clearly identified as an essential component of both risk assessment and risk management.

The four steps draw very heavily upon previous data, expertise, research, and experimentation, especially toxicology (left panel). The characterization of the risks and especially the interpretation of any adverse effects will require involvement of risk management components such as risk perception, social values, economics, policy, and enforcement (right panel), as well as excellent practice in risk communication throughout step 4 and all of risk management (see Chap. 6).

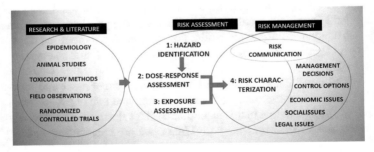

Fig. 3.1   The modified four-step risk assessment model

3.2   Hazard Identification (Step 1)

3.2.1   Data from the Site

The hazard identification step examines the chemical substances or other agents present at the site and asks: *"Which of these substances could present a danger to people who may be exposed under various scenarios, and* via *various realistic pathways?"*

Not only the substances but their concentrations are important. Although there are a few exceptions, extremely low concentrations of even dangerous toxicants are relatively harmless, while even essential nutrients can be harmful in excessive amounts or under certain conditions (even oxygen, iron, sodium chloride, and some vitamins).

A large-scale accidental release of industrial toxic waste onto farmland could involve one or a number of substances contaminating food crops, animal feed, and pastureland. These substances may later appear in meats, grains, produce, dairy products, eggs, or well water, or they may be inhaled directly as airborne contamination or as inhaled dust over time. In step 3 (pathway assessment, Sect. 3.4), we will consider these scenarios and many others including gardening, fishing, hiking, and children playing.

Initial data gathered from the site of the spill, rupture, release, leak, or accident, should include the chemical species, and the concentrations of each substance in the air, the water supply, the food, or any other medium through which it could reach people. (Additional important criteria will include the physical form, such as aqueous solution, gas, solid, liquid, volatility, storage, or transport temperature, whether under pressure or not, and the type of container).

Figure 3.2 lists typical lab results in the form of mean and maximum concentrations in air, groundwater, and soil at a hypothetical site.[2] Note that in this example, "soil" is the medium listed, *not* vegetables grown *in* the soil. Unless we are considering *"pica"*

---

[2]Both systems of scientific notation used in risk assessment, and you will find both in this text. Thus: **2.10E−03 = 2.1 × 10⁻³ = 0.0021;** **1.5E+00 = 1.5 × 10⁰ = 1.5; and 4.2E+02 = 4.2 × 10² = 420.0.**

Chemical	Air		Ground-water		Soil	
	Mean (mg/m³)	Max (mg/m3)	Mean (mg/L)	Max (mg/L)	Mean (mg/kg)	Max (mg/kg)
Chlorobenzene	4.09E−08	8.09E−08	2.50E−04	1.10E−01	1.39E+00	6.40E+00
Chloroform	1.12E−12	3.12E−12	4.30E−04	7.65E−03	1.12E+00	4.10E+00
Heptachlor	1.20E−08	1.44E−08	2.10E−04	4.00E−03	3.00E+00	4.25E+00
DEHP	3.29E−07	8.29E−07	1.80E+00	1.75E−01	1.03E+02	2.30E+02
Gamma HCH	ND	ND	1.10E−01	1.70E−02	1.50E+00	2.60E+00
ND= not detected						

Fig. 3.2 Concentration of chemicals found at the site in air, groundwater, and soil

syndrome, ingestion of soil is not the target, so for intake calculations, the conversion is accomplished by means of a bioconcentration factor (BCF) to convert from soil concentration to the concentration in edible vegetables (see Sect. 3.4.6 and Fig. 3.18) Alternatively, one could also measure the concentrations of contaminants in the vegetables grown in the soil.

3.2.2 Assessing Carcinogens Versus Noncarcinogens

A distinction is needed between exposure to substances that could result in carcinogenic outcomes and substances that could result in "noncarcinogenic" outcomes (which include general and local toxicity, reproductive effects, and neurological disorders). The US Environmental Protection Agency's Integrated Risk Information System (IRIS) database indicates if the substance has "carcinogenic" or "noncarcinogenic" criteria. Many substances that we will use in the examples exhibit both carcinogenic and noncarcinogenic properties. For instance, the mycotoxin *aflatoxin*, elaborated by the mold *Aspergillus flavus* in damp peanuts, can have acute toxic effects on the liver if large quantities are consumed over a short time but can also increase the lifetime risk of cancer of several organs if ingested in small quantities over a longer time. In such cases, the same substance appears in both carcinogenic and noncarcinogenic lists.

3.2.3   The US EPA IRIS Database

Figure 3.3 is a brief, simplified excerpt from the US Environmental Protection Agency's IRIS database that is available online in a comprehensive and informative format (EPA (US) 2022).

Note: The Reference Dose (RfD) and Slope Factor (SF) values shown in Fig. 3.3, and in worked examples throughout the book, were taken from an earlier release of the IRIS dataset. They are for demonstration purposes only and may now be obsolete. For actual application, always check with the IRIS database for the most current values.

The fields shown in the IRIS table are explained below:

- **Chemical name:** Many chemicals are listed under several names, so you may need to look in several places in the dataset. For instance, DEHP is also known as BEHP or by its formal name di(2-ethylhexyl)phthalate.
- **CASRN:** Chemical Abstracts Service Registry Number: Useful to cross-check and verify you have the correct substance.
- **RfD:** The reference dose by which the potency of noncarcinogenic toxicants is measured. Shown here are oral and inhalation RfDs. If an appropriate RfD value is shown, you can proceed with a noncarcinogen assessment. Units are **mg per kg · day.**
- **RfC:** The reference concentration by which the potency of noncarcinogenic toxicants is measured. The units are typically reported in mass per volume (e.g., mg/m^3).

Chemical	CASRN	Oral RfD mg/kg·d	Inhal RfD mg/kg·d	Oral SF 1/mg/kg·d	Inhal SF 1/mg/kg·d	Carc. class.
Toluene	108-88-3	2.0CE−01	1.40E+00	No data	No data	D
Chlorobenzene	108-90-7	2.0CE−02	No data	No data	No data	D
Chloroform	67-66-3	1.0CE−02	No data	6.10E−03	8.10E−02	B2
Heptachlor	76-44-8	5.0CE−04	No data	4.50E+00	4.50E+00	B2
DEHP	117-81-7	2.0CE−02	No data	1.40E−02	No data	B2
Gamma HCH	58-89-9	3.0CE−04	No data	No data	No data	-
n-Hexane	110-54-3	No data	5.72E−02	No data	No data	-

Fig. 3.3 The US EPA IRIS Database (Integrated Risk Information System): Excerpt

- **SF:** Slope factor (also known as **carcinogen potency factor (CPF)** or Q1*). These are taken from dose-response curves and measure the potential of carcinogens. Oral and inhalation SFs are shown. If an appropriate SF value is shown, you can proceed with a carcinogenic risk assessment. Units are **1/(mg per kg · day)** or **(mg per kg · d)$^{-1}$**.
- **Carcinogen classification:** The classification system (Fig. 3.4) employs weight of evidence that the substance is a human carcinogen.[3]

Figure 3.5 shows the list of chemical substances from Fig. 3.2 together with the groundwater concentrations, and the oral RfD and oral SF values from the EPA/IRIS dataset in preparation for an assessment of risk from ingested well water. Note that three of

Code	Description
A	Human carcinogen
B1	Probable human carcinogen (limited human data available)
B2	Probable human carcinogen (sufficient evidence in animals, inadequate/no evidence in humans
C	Possible human carcinogen
D	Not classifiable as to human carcinogenicity
E	Evidence of non-carcinogenicity for humans

(EPA(US)h (1986)

Fig. 3.4 Early (1986) EPA: Classification system for carcinogenicity

| Chemical | Cmax | (EPA) Oral RfD | (EPA) Oral | Categorized | |
| | | (Reference dose) | (Slope factor) | | |
	(mg/kg)	mg/kg·d	1/mg/kg·d		
Chlorobenzene	1.10E−02	2.00E−02	No data	Non-Carc	
Chloroform	7.65E−03	1.00E−02	6.10E−03	Non-Carc	Carc
Heptachlor	4.00E−03	5.00E−04	4.50E+00	Non-Carc	Carc
DEHP	1.75E−01	2.00E−02	1.40E−02	Non-Carc	Carc
Gamma HCH	1.70E−02	3.00E−04	No data	Non-Carc	

Fig. 3.5 Maximum concentrations for five substances found in water samples

---

[3]This system was used by EPA 1986 to 1996, and is still quoted. However, the EPA 2005 classification system (Fig. 3.14) relies more heavily upon the "weight of evidence" narrative rather than the descriptor alone. Section 3.6.1 offers criticism in the context of risk characterization (EPA (US) b).

the substances have *both* carcinogenic and noncarcinogenic toxicity. Thus, three contaminants will be assessed for both carcinogenic and noncarcinogenic risk, while two will only be assessed for noncarcinogenic risk.

3.2.4   Changing Chemicals Over Time

A leaking toxic waste site is often found to contain different chemical species than were originally present. Those chemicals may have undergone oxidative or reductive processes and possibly have been dissociated and recombined over time. An historical spill of DDT could contain lower concentrations of the original DDT and may also include metabolites such as DDE and DDD. Chemicals spilled several decades ago may now only be present at low concentrations, while new species could have appeared. Therefore, it is also important to consider the metabolites of chemicals, not only the parent compound.

3.2.5   Toxicity Scores

When a very large number of chemical species are present at one site, risk assessors have found it useful to identify which should be assessed with higher priority. This involves eliminating chemicals that do not represent a threat because of either extremely low concentrations at the site or extremely low toxic potentials (or both). A **toxicity score** (TS) can be generated for the contaminants (Figs. 3.6 and 3.7). It is important to understand that this is a very quick assessment based simply on two factors: the quantity

Chemical	C_{MAX} (mg/kg)	Ora Slope Factor 1/mg/kg·d	Toxicity score	TS percent	TS accumulative percent
Heptachlor	4.00E–03	4.50E+00	0.018000	87.819%	87.819%
DEHP	1.75E–01	1.40E–02	0.002450	11.953%	99.772%
Chloroform	7.65E–03	6.10E–03	0.000047	0.2277%	100.000%
			(0.020497)	(100.0%)	

Fig. 3.6 Toxicity score (ranked) for three carcinogens found in water

Chemical	C_{MAX} (mg/kg)	Oral Reference dose mg/kg·d	Toxicity score	TS percent	TS accumulative percent
Gamma HCH	0.01700	3.00E−04	56.667	71.116%	71.116%
DEHP	0.17500	2.00E−02	8.7500	10.981%	82.098%
Heptachlor	0.00400	5.00E−04	8.0000	10.040%	92.137%
Chlorobenzene	0.11000	2.00E−02	5.5000	9.9025%	99.040%
Chloroform	0.00765	1.00E−02	0.7650	0.9601%	100.00%
			79.682	(100.0%)	

Fig. 3.7 Toxicity score (ranked) for five noncarcinogens found in water

of the chemical found on site and the toxicity of it. No measure of the dosages for any humans who may be exposed to the chemicals are incorporated, nor of course, of any health effects arising from those exposures.

Toxicity scores are not as commonly used as previously; modern data processing systems will automatically prioritize the contaminants. The toxicity score is calculated as follows:

$$\text{For each carcinogenic chemical: } \mathbf{TS} = \mathbf{C_{max}} \times \mathbf{SF}$$

$$\text{For each noncarcinogenic chemical: } \mathbf{TS} = \frac{\mathbf{C_{max}}}{\mathbf{RfD}}$$

Figure 3.6 shows the three carcinogens from Fig. 3.5 ranked by the toxicity score. The first two chemicals combined account for 99.77 percent of the total carcinogenic potential, while everything else (only chloroform in this example) represents 0.23% and might be removed from the analysis.

Similarly, Fig. 3.7 shows the five noncarcinogens ranked by toxicity score. The most dangerous material present is Gamma HCH, which accounts for 71 percent of the total toxicity. The first four, Gamma HCH, DEHP, heptachlor, and chlorobenzene, when combined, account for more than 99 percent of the total toxicity. Only one (again chloroform) represents less than 1 percent. In keeping with the method of prioritizing by toxicity score, only the first four would be assessed.

Note that the TS calculations here were carried out using C_{MAX}, the "maximum" concentrations found at the site. The "maximum" (usually the upper 95% confidence limit) has been preferred in the past, but increasingly the "mean" is encouraged, as it reduces the excessive error obtained from repeated use of "worst-case" values. It remains important to explain which value you have used.

The ranking of toxicity scores indicates which chemicals pose the greatest hazard based solely on their maximum concentration at the site, and the listed overall toxicity (RfD) and carcinogenicity (SF).

Where used, the TS analysis is intended to identify the substances accounting for 99% of the overall risk, and in so doing, it is not uncommon to reduce 100 detected chemicals, most of which present negligible risk, to 10–15 chemicals for which assessment is more urgently needed.

3.3 Dose-Response Assessment (Step Two)

3.3.1 Limited Human Data

Although the objective is to assess human health risks, accurate epidemiological data are often not available for potentially harmful substances. Aside from a few carefully designed randomized controlled trials usually at very low exposure levels, and a few accidental exposures ("natural experiments"), which can provide valuable data on high-level exposure, most experimentation on human beings is not ethically acceptable. If no human data are available, living animal (in vivo) or laboratory cellular (in vitro) data are used. Extrapolation from cells or animals to human systems is not always reliable, and for this reason, regulatory bodies design exposure guidelines using uncertainty factors to account for this.

3.3.2 Carcinogens and Noncarcinogens

Carcinogens often have a high profile in popular anxiety and are associated with fear and dread. This can be attributed in part to the typically long intervals between exposure and the appearance of the disease, the lack of "knowing" whether an exposure has even taken place, and a general sense of powerlessness to reverse or change the outcome even when the diagnosis has been made. Assessing the risks for carcinogens also requires a different approach to assessing risks for noncarcinogens. Unlike most toxicants, where we can be confident that a specific dosage will prove lethal to a certain mammal or insect, carcinogens are best described as following a "probabilistic" incidence model. Most people know of a person who has smoked 25 cigarettes a day for 75 years and has not developed cancer. These types of examples are not common, but they exist. Cancer is typically predicted as a probability rather than a certainty.

3.3.3   Three-Step Model of Carcinogenesis

An *initiator* can be a chemical, radiological, or mechanical intervention that damages a part of the genetic code in the nucleus of the cell, causing insertion, deletion, or substitution of genes. Examples of initiators (*genotoxic carcinogens*) are benzene, aflatoxins, benzo-a-pyrene, acetaldehyde, nitrosamines, x-rays, and asbestos fibers. Most of the time, this damage renders the cell incapable of reproduction (and the natural repair processes also help to ensure the cell cannot cause trouble). However, on probability alone, some mutations may cause the cell to reproduce out of control.

However, damaged DNA is not sufficient. The cell must undergo reproduction (through mitosis) for the damaged genes to cause a tumor. The cell's reproduction may be due to genetic programming to replace tissue, or the cell may be stimulated to repair damaged tissue following disease or injury. Any mechanism causing the cell to reproduce is considered a *promotor* (a *non-genotoxic carcinogen*), and a wide range of substances and mechanisms can act as a promotor, including salt, alcohol, or other irritants, as well as any physical damage such as stomach ulcer or surgery, chemical damage such as gastroesophageal reflex disease (GERD), or any other inflammation that stimulates cellular repair. The final step of this carcinogenesis model is *progression*. In this phase, cells compete with each other to survive, often leading to

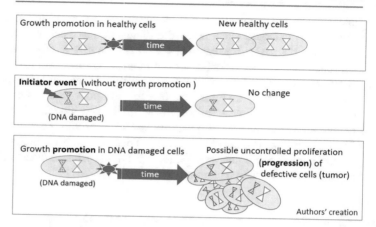

Fig. 3.8 Three-step model of carcinogenesis

further mutations and more aggressive tumor growth. Figure 3.8 illustrates the three-step model of carcinogenesis as compared to healthy cellular growth.

It is worth noting that when animal studies are carried out on food additives or pharmaceuticals for carcinogenicity, the dosages given to the rodents are initially far in excess of their normal intake. Even if the substance is not an initiator carcinogen, the very high, almost-toxic concentration in the animal's tissues, especially liver, stomach, intestines, kidney, and bladder, can cause inflammation in those organs, which acts as a promotor carcinogen. Clearly, then, any tumor-causing processes have to be carefully assessed to determine if the substance was really capable of causing genetic damage (an initiator), or if the tumor was only due to inflammation from the maximally tolerated dose (MTD), in which case it is a promotor.

3.3.4 Threshold Models

Risk assessment for carcinogens and noncarcinogens also differ in terms of the concept of "threshold" and how it is applied. A *noncarcinogen* has a concentration (a "threshold" dose) below

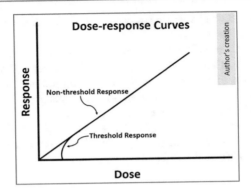

Fig. 3.9 Threshold vs. non-threshold curves

which no effects can be detected on the cellular, subcellular, or molecular level. Noncarcinogenic dosage below the threshold is a "safe" zone, and dose-response curves only come into the picture at or above that level (Fig. 3.9). Lead is an important exception (see the note below).

Most repair mechanisms (e.g., enzymes) exist with a large number of identical copies. The destruction of a few of these will have little effect on the organism. Only when a significant fraction of the targets has been eliminated by toxic action, that is, above a certain threshold for the target dose, will a toxic effect occur.

Carcinogens, on the other hand, at least in conventional risk assessment models, are not considered to have threshold levels, and the pathogenic effects are theorized to begin at any exposure. This is called the *linear, no-threshold model* (LNTM) for carcinogens, and it is still considered a sound "conservative" basis for carcinogenic risk assessment, even though there is some empirical evidence that such thresholds *do* exist for carcinogens.

Some materials deserve special comment:

- *Carbon tetrachloride:* Intake of CCl_4 results in pathologic damage to liver tissue; however, the liver will continue to function normally, eventually replacing the damaged cells. At some threshold, the liver will become dysfunctional, and the damage may not be reversible.

- **Lead:** Conventional wisdom previously claimed that while blood lead levels (BLL) are above 10 µg per 100 mL, blood inhibited cognitive development in children and BLLs below this threshold would not harm the developing brain. More recently, however, this has been disputed, and several authorities advance the principle that **no** threshold level of lead is "safe" for embryo development or children of any age (Health Canada 2013; WHO 2022).

3.3.5 Target Organs and Systems

Noncarcinogenic effects include a very wide range of mechanisms that result in tissue damage, loss of function, organ failure, and death. Some directly damage cell components, and others interfere with enzyme systems, transport mechanisms, or metabolic pathways. The majority of toxicants disrupt or block the effects of enzymes. In recent years, a new concern has arisen regarding hormonal disruptive effects leading to reproductive and fertility problems, low birth weight, and preterm delivery.

3.3.6 "Dose" Versus "Dosage"

The *"dose"* is understood to mean the actual quantity taken in by an individual (animal or human), for example, 5 mg on a single occasion, or 15 µg per day. Far more valuable in toxicology and risk assessment work is the *"dosage,"* which standardizes the intake of the substance in terms of *"per kg body weight."* In this way, we can compare the intake of a substance as 2 mg per kilogram body weight whether the individual was 90 kg or a 16 kg child. With great caution, we can even extrapolate dosage in mg/ kg (body weight) from laboratory test animals to humans.

3.3.7   NOAEL/LOAEL

These are common ways of linking adverse or deleterious effects to specific dosages, based on the accumulated evidence from both experimental (usually animal) and observational studies.

Of particular interest is the lowest intake dosage which has been associated with adverse (toxicological) effects (LOAEL), and the highest intake dosage associated with no adverse (toxicological) effects (NOAEL). These are shown in Fig. 3.10.

In addition, the *no observed effects level* (NOEL) is a version of NOAEL recording the highest dosage without causing <u>any</u> effect. And the *lowest observed effect level* (LOEL) is the lowest dosage tested, for which <u>any</u> effects were observed.

Limitations with LOAEL and NOAEL When citing the LOAEL, NOAEL, etc., it is important to be aware of their limitations. They are simply the lowest and highest dosages known to be associated with adverse effects and no adverse effects, respectively, but the next peer-reviewed study may change that level, especially if this area has not been well studied at least at these dosage levels. The studies that inform these levels may have used

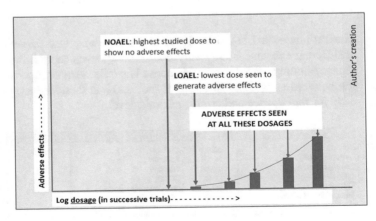

Fig. 3.10   Relative positions of NOAEL and LOAEL

insufficient numbers of animals, or for periods of time inadequate to show effects. Statistical uncertainty can arise from inappropriate use of the standard deviation and standard error where the majority of toxic effects follow a log-normal distribution, so symmetry cannot be assumed around a mean. And finally, the NOAEL (and the LD_{50}) gives no indication of individual variations in susceptibility.

The dose-response curve often indicates the degree of variability: a steep curve (e.g., cyanide) generally means a strong response, with individual variability insignificant within the main effect of the chemical. An almost flat or weak curve (e.g., ethylene glycol) indicates large variation between individuals, thus detracting from the confidence of the NOEL.

3.3.8 LD50

The dosage at which 50 percent of the test animals die is called the LD_{50} (lethal dose 50%). The measure is commonly used to compare potency between toxic materials (as is also the LC_{50}, the concentration of vapor or gas that is lethal to 50 percent of test animals). Figure 3.11 shows a selection of LD_{50} values for common substances. The lower the LD_{50}, the more dangerous the chemical is.

Limitations with LD_{50} The slopes of the dose-response curves of different substances may vary considerably, even though the dosage eliciting the 50 percent response is similar. Any comparison between substances in terms of their LD_{50} is therefore *only valid for that dosage* and no other dosage level.

Substance	LD50 (mg/kg)	Substance	LD50 (mg/kg)
Ethyl alcohol	10,000	DDT	100
Sodium chloride	4,000	Paraquat	4
Ferrous sulphate	1,500	Nicotine	1
Chlorobenzilate	960	(TCDD)	0.001
Chlordane	457-590	Botulinum neurotoxin	0.00001
Sodium phenobarbital	150		

Fig. 3.11 Acute LD_{50} for several chemical compounds

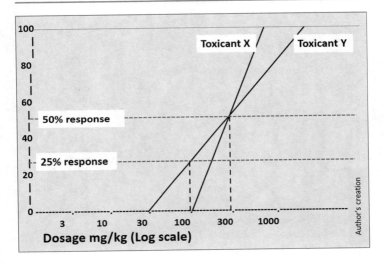

Fig. 3.12 Comparing LD$_{50}$ values

Figure 3.12 demonstrates the same LD$_{50}$ dosage (300 mg/kg) for two hypothetical substances, X and Y. (Note that the dosage scale is logarithmic, producing a nearly straight dose-response curve.)

The LD$_{50}$ dose is the same (300 mg/kg), but at all other dosages, the overall dose-response is markedly different. At around 100 mg/kg, Y has already killed 25 percent of the test animals, whereas X has produced no effect at all. And at approximately 600 mg/kg, toxicant X has a lethality rate of 100% while toxicant Y has a lethality rate of approximately 70%. The LD$_{50}$ alone would fail to identify X as the far more dangerous toxicant.

3.3.9   The Slope Factor for Carcinogens

"Potency" is generally considered a measure of the probability of a substance to induce cancer over a lifetime for a given dosage. The key factor in calculating risk is the "slope factor" (SF), also known as the "carcinogenic potency factor" (CPF) or Q1* (pronounced "Q one star").

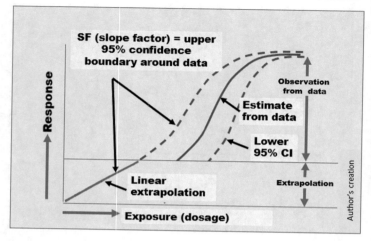

Fig. 3.13 Slope factor (SF)

The SF is calculated as the upper 95% confidence boundary around the dose-response curve (Fig. 3.13).

This is sometimes incorrectly written as the "maximum" concentration although a small number of observations can still theoretically exceed the upper $CI_{95\%}$ boundary.

You will notice that the units for the SF (e.g., in the IRIS database) are expressed $(mg/kg\cdot day)^{-1}$ or $1/(mg/kg\cdot day)$, which is the inverse of the chronic daily intake $(mg/kg\cdot day)$. The rationale will be clear later when we calculate and characterize the risk in step 4 by multiplying the slope factor by the chronic daily intake. The units will cancel completely, leaving the unitless probability (risk) of dying of cancer for the given exposure.

The US EPA classifies substances as *known, probable*, or *possible human carcinogens* on the basis of the available epidemiologic association with human cancer and the accumulated evidence of the ability to induce cancer in multiple species of test animals. The previous six-point system (see Fig. 3.4) was used by EPA from 1986 to 1996 and is still often quoted.

The EPA 2005 classification system (Fig. 3.14) incorporates a summary of both experimental and epidemiological information

USEPA 2005 weight-of-evidence classification system for carcinogenicity
• Carcinogenic to humans (e.g., causal)
• Likely to be carcinogenic to humans
• Suggestive evidence of carcinogenic potential
• Inadequate information to assess
• Not likely to be carcinogenic to humans
Ref: EPAd 2005

Fig. 3.14 USEPA 2005 weight-of-evidence system

and is also an indication of the relative weight that each source of information carries (human and/or animal experiments). In brief, it relies more heavily upon the "weight-of-evidence" narrative than the A-E ranking system alone (EPAd 2005).

3.3.10 Special Note About "Inhalation" and 'Fugitive Dust"

The examples and calculations employed in this book for inhalation are limited to the intake of vapors and gases. A separate process for determining pollutants that are *absorbed* and especially *adsorbed* onto inert solid particles is not included here. Inhalable (respirable) particles are usually carried in exhaust gases and other effluvia from industrial or incineration operations.

Total suspended particulates (TSPs) are monitored and published for many urban and some rural areas in North America. Those with less than 10 μm diameter (PM_{10}) are of greatest concern, with particles 2.5 μm ($PM_{2.5}$) especially capable of being deposited in the gas-exchange region (alveoli) of the lungs.

3.3.11 Sources of Toxicological Data

The risk analyst has standard sources of toxicological data available such as the *Agency for Toxic Substances and Disease Registry* (ATSDR) and the EPA's *Integrated Risk Information System*

(IRIS). The information is updated monthly and available online. IRIS contains both qualitative and quantitative data, for both carcinogens and noncarcinogens (EPA(US)e 2011). Some of the values for SF and RfD used in the worked examples in this book may be obsolete, so for current values, please refer to the constantly updated IRIS data online at https://www.epa.gov/iris. In Canada, toxicological reference values (TRVs) can be obtained from Health Canada (updated 2013) or provincial bodies.

Based upon the best available information to date, many of the toxicological indices used in calculations of carcinogenic and noncarcinogenic risks include different kinds of variation. "True" variability is due to known variation in values, while uncertainty can be due to insufficient or unknown data, flawed data collection, or confounding. In addition, safety factors are purposefully built in to reference doses (RfDs) and carcinogenic slope factors (SFs). These are systematic errors invariably made in the direction of *protecting* public health by ensuring that risks are *over*estimated rather than *under*estimated.[4] Section 3.6.1 discusses this from a critical perspective and describes intake estimates constructed from models where each variable is assumed to be excessively high. Additional examples are as follows:

- For carcinogens, the multistage model assumes the upperbound 95% confidence limit from extrapolated data. This means theoretically that fewer than 2.5 percent of the population would be expected to have a higher risk than the estimate from that one factor.
- For carcinogens, the multistage model extrapolates data from the 10–90% carcinogenesis range observed in experimental

---

[4]A *"conservative estimate"* in normal speech usually denotes a value slightly less than the most likely. In risk assessment, *"conservative estimates"* are generally overstated values, such that the final calculation of risk will be closer to the "worst-case" outcome. The rationale for this is that a model based upon these higher values will include a larger proportion of the population, with a very small percentage of people remaining who would actually still exceed this risk assessment. This is doubtless one of the many sources of confused communication about risk.

animals to the regulatory target of 0.0001% carcinogenesis, a set which could overstate risk by several orders of magnitude.

- Although evidence shows that, like noncarcinogens, non-genotoxic carcinogens (promotors) have thresholds below that they fail to influence cellular division, they are nevertheless treated mathematically like genotoxic (initiator) carcinogens according to the model.

3.3.12 Safety Factors for Noncarcinogens (ADI, RfD)

- **Acceptable daily intake (ADI)**: The daily intake of food item, ingredient, or additive that should not produce an adverse health effect. ADIs are based upon NOAELs but are reduced by safety factors that allow for individual variability. For instance, an uncertainty factor of 10 is generally used to allow for variation between individuals, and an additional factor of 10 allows for extrapolation from experimental animals to humans.
- **Tolerable daily intake (TDI):** An estimate of the amount of a substance in food or drinking water that is not added deliberately (e.g., contaminants) and that can be consumed over a lifetime without presenting an appreciable risk to health
- **Reference factor dose (RfD)**: A version of ADI reduced by several additional factors, for example, by 10 if data are derived from sub-chronic instead of chronic study, and by a further 10 if data are extrapolated from LOAEL to NOAEL, and for animal-human extrapolation.
- **Reference factor concentration (RfC)** is derived in a similar way to RfD but uses contaminant concentration instead of dose.
- **Inhalation unit risk (IUR):** An estimate of increased carcinogenic risk due to inhaled contaminants to a concentration of $1\ \mu g/m^3$ for a lifetime.

  The relationship between these dosage values is depicted in Fig. 3.15.

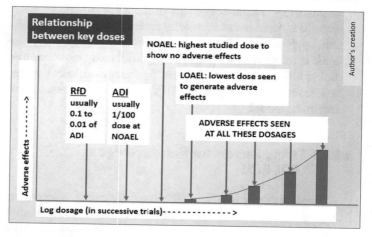

Fig. 3.15 Key dose values relative to NOAEL/LOAEL

3.4   Exposure Assessment (Step 3)

At this stage, we are asking "What are the important characteristics of those who are exposed? For how long, and how often? Can we determine their exposure dose or dosage?"

3.4.1   Components of the Pathway

This is the chain of events starting with the accidental (or intended) release, leak, or use of a chemical in the environmental or worksite, and leading to that chemical reaching workers or other people (the "receptors") in sufficient quantities to represent a risk to health or life.

It is important to determine what happens to these chemicals: how they are transported, transferred, or transformed and in what medium (soil, water, air, in food, dust, etc.). A great number of mechanisms could act to transfer the contaminants to another medium or to storage sites (e.g., sorption onto soils).

A *pathway* will normally involve the following components:

- **Source:** For example, a storage tank, cylinder, reactor vessel, waste treatment lagoon
- **Chemical release:** Mechanism (e.g., defective valve, ruptured pipe, dropped bottle, spills, spraying, erosion, volatilization, fugitive dust generation, leachate)
- **Transport:** Mechanisms (e.g., natural air flow, ventilation systems, dust)
- **Transfer:** Mechanisms (e.g., adsorption, absorption)
- **Transformation:** Mechanisms (e.g., neutralization, biodegradation). Some degradation products are less toxic as they become transformed into simpler, more oxidized substances. In other cases, the degradation products may be *more* toxic than the original.
- **Exposure point:** The point of contact between the receptors (people) and the substance. This might be air, exposed skin, contaminated food, or water.
- **Receptors:** People (e.g., workers, bystanders, family/community members)
- **Exposure route:** Into the body (e.g., ingestion, inhalation, transdermal absorption)

An example demonstrating the applications of these terms is offered as Fig. 3.16.

Transformation can present a complicated scenario. With the passage of time, the "pool" of chemicals present at a site may not resemble the original contaminants. Some toxicants may oxidize to simpler compounds, for example, deadly hydrogen cyanide (HCN) becomes CO_2, oxides of nitrogen and water, but in other examples, the degradation products may be *more* toxic than the original chemical (e.g., when polyvinyl chloride is incinerated at an inadequate temperature.

Those assessing risks should also be aware of the common situation in which multiple pathways can develop, some of which may be missed in the initial assessment. Although this is especially common in an agricultural environmental (see Case Study #20), it can also be seen in worksite-related risk assessment, espe-

AMMONIA AT DAWN

Shortly before dawn on a rainy November morning, a vehicle transporting cylinders of ammonia gas [source] collided with a fallen tree. Two cylinders were ruptured, [releasing] a cloud of ammonia gas.

A slight breeze [transport] began to move the gas into the direction of a farming community 1 Km to the east. In the twenty minutes between the release and the gas reaching the village, more than 80% of the ammonia had been absorbed [transfer] by the light rain, producing a weak solution of ammonium hydroxide [transform].

Some residents [receptors] inhaling the ammonia [exposure point] suffered respiratory distress. All recovered, although several elderly residents complained that their asthma had become much worse.

During the following week, several residents complained of ammonia taste when drinking [exposure point] their well water [exposure route], although the public health laboratory was unable to confirm the presence of ammonium hydroxide in well water samples.

During the following week, surface water monitoring revealed that all traces of the ammonium hydroxide in ditches and streams in the area had disappeared, due to the acidic nature of the soils and clays in the area [transform]. Groundwater testing continued for a month but with no unusual findings.

Fig. 3.16 Identifying pathway components from "source" to "receptor"

cially when contaminated clothing is caried off-site to the family environment.

In addition to the identification of the source, the release mechanism, the transportation, the transfer, and transformation of the chemical, several further details are needed for a comprehensive risk assessment process:

- Identifying exposed populations (...*who?*)
- Development of exposure scenarios (...*how?*)
- Determination of exposure point (...*where?*)
- Determination of concentrations (...*how much exposure?*)
- Estimation of receptor doses (...*how much retained?*)

3.4.2   The Migration/Movement of Substances

The movement of substances away from the original worksite is an extremely important consideration, and numerous cases have been examined where lead, asbestos, and other dangerous materials are carried on shoes or clothing, reaching the family of the worker or other community members. Overalls from work brought home for washing can introduce toxic dusts into the household, exposing its occupants (for potentially long-term exposure), including pregnant women, infants, and children who are particularly at risk of permanent developmental damage when exposed to heavy metals.

The movement of materials over a wider area becomes particularly important when a risk is large scale and poorly contained. Nine million residents in Michigan consumed contaminated meat and milk during the mid-1970s after PBB (polybrominated biphenyl) – a toxic fire retardant – was erroneously added to dairy cattle feed and distributed to farms throughout the state, resulting in contamination of pasture, livestock, meat, dairy products, and crops grown on manured land, for several years.

3.4.3 Identifying Exposed Individuals

The "recipient" group is often larger than originally supposed, and scenarios should be carefully set up to include all the potential groups.

- Workers using the material
- Others using the same equipment or worksite
- Other workers in the area at the time or at any time afterward
- Clean-up workers in the event of a spill or release
- Those in the community who might use the product
- Those in the community (especially household members) who might be exposed to the worker's clothing and equipment
- Transportation staff (drivers, pilots, loaders, baggers, etc.)
- Children at play – field, playground, backyard, etc.
- Residents working in their garden
- Joggers, walkers, swimmers, hikers
- Illegal, such as trespassers

3.4.4   Development of Exposure Scenarios

Each scenario sets out a set of conditions, characteristics, and activities that identify a specific group of workers or other members of an exposed population. They should not be limited to formal, legitimately defined groups but include trespassers or children who enter a transformer compound to collect a ball.

The following scenarios are presented together with some relevant questions for the assessor to consider when determining whether the scenario warrants evaluation (Fig. 3.17).

Worker scenario	Could workers be exposed because of a change in the use of a site or due to involvement in remedial activities? Noted that workers are generally protected under separate OHS provisions. Potential risks for workers are usually estimated to determine the potential for *unacceptable* risks, rather than as a mandated exercise designed to ensure that workers are not placed at risk during the normal course of their activities.
Trespass Scenario	What evidence is provided that trespassing may occur at the site? Are there barriers to limit trespass? Are these intact? Are there items on site that may seem inviting to strangers, including children?
Construction scenario	Will construction result in potential exposures for both on-site receptors (e.g., direct contact with soils by construction workers), and off-site populations (e.g., nearby residents' exposure to fugitive dusts released during the work, or workers' families' exposure to laundry, etc.)?
Residential use scenario	Is the area used by local residents now or in the future? Can exposure pathways affect the local residents (e.g., air or water)? A residential scenario is frequently evaluated as a hypothetical condition in an effort to estimate worst-case risks. Exposure under a residential scenario generally represents the greatest exposures and the highest potential risks.
Recreational water-use scenario	This scenario is particularly useful in evaluating potential risks associated with swimming, canoeing, wading, fishing, etc. It may be useful to confirm the regulated use and classification of the water body with the appropriate agency.

Fig. 3.17 Examples of exposure scenarios

3.4.5   Exposure Point Concentrations

The risk assessor or their team should determine the concentrations of contaminants at all exposure points for all pathways. This includes ground and surface water, soil and sediments, and food (grown, foraged, and hunted or fished). Future conditions may vary. For example, present contamination may change such as a plume. Thus, it is important to not only assess baseline conditions but anticipate future changes. This would involve the use of fate and transport modelling methods (US EPA 2016).

3.4.6 Bioconcentration Factor (BCF)

Where data are only available for the "medium' (e.g., the soil or the water), we may need to extrapolate to obtain the concentration that we would expect in the vegetables (grown in that soil) or in the fish (from the lake or river water). Some BCF values are available to aid in this conversion but largely depend on the contaminant's chemistry (Fig. 3.18).

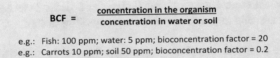

$$BCF = \frac{\text{concentration in the organism}}{\text{concentration in water or soil}}$$

e.g.: Fish: 100 ppm; water: 5 ppm; bioconcentration factor = 20
e.g.: Carrots 10 ppm; soil 50 ppm; bioconcentration factor = 0.2

Note: this is a single step, and not the same as *biomagnification*

Fig. 3.18 Bioconcentration factor

3.4.7 Incorporating Other Models

For groundwater contaminants, hydrogeologic models can be used to estimate the future concentration at a downstream well. For volatile organic compounds released to the atmosphere, a Gaussian diffusion model can be employed to estimate downwind concentrations. You may need the help of experts with experience in standard plume dispersal formulae, atmospheric mixing models, and adiabatic rates. In general, the level of effort employed in data collection and modelling will depend upon the estimated complexity and severity of the risk. Nominal risks do not warrant the same level of analysis as the clearly significant risks. All mathematical models involve assumptions, but it is essential that the appropriateness of these assumptions be reviewed and always stated clearly.

In brief, exposure assessment describes the magnitude, duration, and route of exposure, which characterizes the exposed populations by age, sex, race, and size. It should also address the **uncertainties** in all the estimates described. Exposure assessment can sometimes be used to comment on or predict the feasibility of regulatory control options and the effect of control technologies on reducing exposure.

3.4.8 Dose Categories

The final step in the exposure assessment stage is to estimate the intake doses of the different chemicals to which the receptors are potentially exposed at the exposure points. Three exposure points are considered: ingestion, inhalation, and dermal contact. The doses are also measured in different ways (Fig. 3.19).

- The *administered dose* (the amount ingested or inhaled): This is the usual value used in estimating the dose to the "receptor." It is essentially the amount taken into the lungs through respired air, or into the alimentary canal through food or water, but with no allowance for any amount that is *not* absorbed. Thus, any of the chemical (e.g., a solvent) that is exhaled or excreted is not deducted from the initial intake.

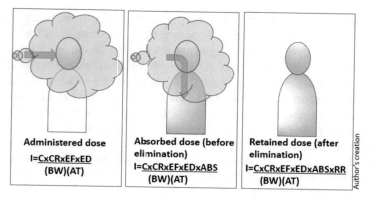

Administered dose

$$I = \frac{C \times CR \times EF \times ED}{(BW)(AT)}$$

Absorbed dose (before elimination)

$$I = \frac{C \times CR \times EF \times ED \times ABS}{(BW)(AT)}$$

Retained dose (after elimination)

$$I = \frac{C \times CR \times EF \times ED \times ABS \times RR}{(BW)(AT)}$$

Author's creation

Fig. 3.19 Types of dosage used in intake measurements

- The *absorbed dose* (the amount absorbed by the body): This is the amount of substance taken into the body tissues (i.e., the administered dose minus the quantity not absorbed). When the proportion absorbed is not available, the absorption is assumed to be 100% and calculated as 1.00, making it the same as the *administered dose*.

- The *retained dose* (the amount that reaches the target organ system): This is rarely used in risk assessment but is used in toxicology, pharmacology, etc. When the proportion retained is not available, the retained amount is assumed to be 1.0. Even when the retained dose can be calculated, one of the criticisms of modern risk assessment is that it assumes an oversimplified single-compartment toxicological model in which all chemicals are absorbed and then evenly distributed throughout the body with no preference for specific tissues.

3.4.9 Calculating the Chronic Daily Intake (CDI)

The following general equation is used to find the chronic daily intake (**CDI** or simply **I**):

$$I_{C \, or \, N} = \frac{\left(C \times CR \times EF \times ED \times ABS \times RR\right)}{\left(BW \times AT\right)}$$

where:

- I_C = intake in (mg/kg(body weight) · day) for carcinogens
- I_N = intake in (mg/kg(body weight) · day) for noncarcinogens
- **C** = concentration at exposure point (mg/L in water, mg/m^3 in air, mg/kg food)
- **CR** = daily contact rate for the medium (L/day (water), m^3/day (air), or kg$_{(food)}$/day)
- **EF** = frequency (days/year)
- **ED** = exposure duration (year)
- **BW** = body weight (kg)
- **AT** = averaging time (days) [see note below]

- **ABS** = absorption rate; assume 1.0 (100%) if not known
- **RR** = retention rate; assume 1.0 (100%) if not known

3.4.10 Helpful Hints in Preparing to Calculate Intake

The averaging time (AT) will depend upon the type of threat being evaluated. In the assessment of noncarcinogenic chronic effects, the AT is the total *exposure duration from beginning to end (expressed in days)*. Exposure to *carcinogens*, however, uses an AT that is always a lifetime (assumed to be 70 years or 25,550 days) to be consistent with the approach used to develop slope factors, and the differing nature of carcinogenic risk as compared to the risk from exposure to toxicants. Briefly, the accumulated carcinogenic risk is considered to remain with you for a lifetime, whereas the risk from noncarcinogenic toxicants is taken to decline when exposure is ceased. When assessing carcinogenic risk for a lifetime, this is often referred to as "lifetime average daily dose" or LADD.

Many of the parameters used in this type of calculation (e.g., skin surface area, air breathed, water ingested, etc.) are available in risk assessment literature. Others, such as exposure frequency or duration, may be specific to the site and may require professional judgment. Common sense also plays a role; if an assessment required an estimate of the number of days during which children's bare skin contacted a lawn or garden soil, the value for Fort Erie, Canada, would be quite different to that for Fort Lauderdale, USA! (See Fig. 3.20 for some useful values.) Use these where you don't have any actual measured values from the studied population.

Contact rates (CR) for breathed air is usually given in m^3 per hour and has to be changed to m^3 per day. In a residential scenario, it would be 24 hours per day, although fewer hours in school, workplace, or recreational scenarios. For example, the complete

Abbrev	Parameter	Adult	Child 6-12 y	Child 2-6 y
BW	Av. Body weight (kg)	70	29	16
CR water	All water intake (L/day)	2	2	2
CR air	Inhaled (M³/hr)****	0.83-1.20	0.46-0.75	0.25
RR	Retention rate (if unknown):	1.00	1.00	1.00
ABS	Absorption rate (if unknown):	1.00	1.00	1.00
Soil	Ingested (mg./day)***	100	100	200
Bathing	Duration (minutes)	30	30	30
EF	Exposure frequency (days/y)	Up to 365	Up to 365	Up to 365
ED	Exposure duration (years)	actual	actual	actual
ATc	Averaging time (carc.) (days)	25,550	25,550	25,550
ATN	Averaging time (non-carc.) (days)	Total days**	Total days**	Total days**
	Extracted from Canada Govt. (1995). Similar values are found at EPA(US)e (2011)			

NOTES to accompany anthropometric data table (Fig. 3.20):

* AT for all carcinogens is a lifetime of days (365x70) regardless of length of exposure.

** AT for non-carcinogens is the actual number of days of exposure from day 1 to the end.

*** Soil ingested may increase enormously with 'pica' individuals (Wilson et al. 2013).

**** Air inhaled varies with activity, gender, body mass, and ethnicity.

(In all cases, actual data if available is to be used preferentially in place of tabled estimates.)

Fig. 3.20 Anthropometric table (default) parameters

though lengthy calculation for a respiration rate of 1 m³/h over an 8 h day, keeping all units visible, could be written as:

$$\left(1\,m^3 \, / \, h\right) \times \left(24\,h \, / \, d\right) \times \left(8\,h \, / \, 24\,h\right) = \left(8\,m^3 \, / \, d\right)$$

However, most people would shorten the calculation to:

$$\left(1\,m^3 \, / \, h\right) \times \left(8\,h \, / \, d\right) = \left(8\,m^3 \, / \, d\right)$$

Note that the volume of air breathed per hour is shown in Fig. 3.20 as a range. More strenuous activity, for example, a workplace requiring physical activity, or exercise, would require a greater volume to be used in the calculations. Similarly, if showers were the only method of washing, 30 minutes may be excessive. Water intake for strenuous or hot work (e.g., iron foundry) would be much greater.

Exposure duration (ED) is the actual duration in years, so six months would be entered as 0.5 y.

In some cases where an assessment *up to the present* time is being made for a population with a mixed age range, the median age is considered as 30 years, an "average" ED for the community. The rationale is that if the assessment covers the past up to today, some elderly people may have lived at the location their entire life, while some arrived (or were born) only recently. Thirty years can be considered a "mean" exposure duration for such a mixed group.

Measured observations (empirical data) should always be used if available and should always be preferred to the table of standard values. For instance, in a community where most of the population is from southeast Asia, the mean adult body weight, as shown in anthropometric surveys, could be 55 kg, not 70 kg. The demographic characteristics can vary enormously in age, too. For instance, the median age (half the population is below and half above) in Japan is 47.3 y and in Kenya, the median is 19.7 y [based on Central Intelligence Agency (CIA) and United Nations (UN) estimates, 2022].

3.4.11 Worked Intake Calculations

The following series of examples have been selected to demonstrate these methods. Please note the importance of always showing the correct units for each variable and cancelling them correctly. In this way, you reduce your chance of making a calculation error. The final chronic daily intake (CDI or I) must be in the form shown: mg/kg·day. Anything else, and you may have an error.

Intake Calculation 1: Dichloroethane in Potatoes
Determine the chronic daily intake for a 70 kg farmer growing potatoes for food in soil above an old toxic waste site. The soil contains an average concentration of 0.023 mg/kg dichloroethylene. Assume 0.5 kg potatoes eaten every day for a lifetime. Bioconcentration factor (BCF) is 2.0.

The EPA data determines dichloroethylene to be both carcinogen and noncarcinogen. For lifetime exposures, the (ED×EF) days and (AT) days cancel, so the intake can be simplified to:

$$I = \frac{[C] \times CR}{BW}$$ for *both* carcinogenic and noncarcinogenic
intakes.

$$I = \frac{\left[0.023\,mg/kg\left(potatoes\right)\right] \times 2.0\left(BCF\right) \times 0.5\,kg/d}{70\,kg\left(bw\right)}$$

$I_c = I_N = (0.0003286 \text{ mg/kg·day})$

Intake Calculation 2: Chloroform in Well Water

Calculate the intake for a 70 kg worker who drinks 2 L/d of well water containing 1 ppm (1 mg/L) of chloroform for 10 years. (Assume no other water is used by this person.)

The EPA data determines chloroform to be both carcinogenic (SF) and noncarcinogenic (RfD).

Intake calculations will differ due to the AT.

$$I_c = \frac{[C] \times CR \times EF \times ED}{BW \times AT} = \frac{1.0\,mg/L \times 2\,L/d \times 365\,d/y \times 10\,y}{70\,kg \times 25,550\,d} =$$

$I_c = (0.004082 \text{ mg/kg·day})$

$$I_N = \frac{[C] \times CR \times EF \times ED}{BW \times AT} = \frac{1.0\,mg/L \times 2\,L/d\,365\,d/y \times 10\,y}{70\,kg \times 365\,d/y \times 10\,y}$$

$I_c = (0.02857 \text{ mg/kg·day})$

Intake Calculation 3: Chlorine Dioxide Near a Water Treatment Plant

A resident living next to a municipal water treatment plant is experiencing a persistent chemical smell. Testing shows a steady concentration of 2 μg/m³ (0.002 mg/m³) of chlorine dioxide (ClO_2) at the residence. Assume the sedentary resident lives all the time at the location, and the exposure direction has already been five years.

Solution: The EPA's IRIS database shows chlorine dioxide with an inhalation RfD of 5.72E−05 but no inhalation SF/CPF, so we proceed to assess it as a noncarcinogen.

Intake (I_N) for this adult can be calculated as:

$$I_N = \frac{[C] \times CR \times EF \times ED \times RR \times ABS}{BW \times AT}$$

C = 0.002 mg/m³ (given)
CR = 0.83 m³/h × 24 h/d = 19.92 m³/day (Fig. 3.21)
EF = 365 days/yr. (given)
ED = 5 yrs. (given)
RR and ABS (unknown), assumed both to be = 1.0 (conservative)
BW = 70 kg (Fig. 3.21)
AT = 365 days × 5 yrs.

Thus,

$$IN = \frac{\left[0.002\,\text{mg}/\text{m}^3\right] \times \left(19.92\,\text{m}^3/\text{d}\right) \times \left(365\,\text{d}/\text{y}\right) \times \left(5\,\text{y}\right) \times \left(1.0\right) \times \left(1.0\right)}{70\,\text{kg} \times \left(365\,\text{d}/\text{y} \times 5\,\text{y}\right)}$$

$$\boxed{I_N = 0.00057 \text{ mg/kg·day}}$$

Note: Where the noncarcinogenic exposure is "all day, every day" *for a given period*, the ED and EF values cancel with the AT value. In this case, this intake calculation is simplified to:

$$I_N = \frac{\left[0.002\,mg/\text{m}^3\right] \times \left(19.92\,\text{m}^3/\text{d}\right)}{\left(70\,kg\right)} = 0.00057\,mg/kg \cdot day$$

Intake Calculation 4: TCE in School Well Water
A rural primary school has been found to have *1,1,2,2-tetrachloroethane* (TCE) in the well water at levels close to 6 µg/L. An alternate water source has been provided but the risk accumulated by the students must be calculated.

The first step is the intake calculation. Assume the children are 6 years old and the exposure has been for one school year.

TCE has an oral SF/CPF value and is therefore a carcinogen. Table (Fig. 3.20) shows 6 yr/old children halfway between the age groups, and therefore we take 23 kg body weight (halfway), 2 L water/d, present in school 7 h/24 h, while the EF × ED as simply the total number of days at school in the whole school year: (10 months × 20 days/month − 23 days civic holidays, vacations − 4 days sick) = 173 d/y × 1 y. We can assume RR and ABS to be 100% and convert 6 μg/L to 0.006 mg/L.

$$I_c = \frac{[C] \times (CR) \times (EF) \times (ED) \times (ABS) \times (RR)}{(BW) \times (AT)}$$

$$I_c = \frac{[0.006\,mg/L] \times (2\,L/d) \times (7h/24h) \times (173d)}{(23\,kg) \times (25{,}550\,d)} =$$

$I_C = 1.03 \times 10^{-6}$ mg/kg·day

Intake Calculation 5: Nickel Refinery Dust

Calculate the intake for nickel refinery dust where workers are present for eight-hour shifts, and work 230 days a year. The exposure has been present for 8 years. The nickel dust is present at 0.5 mg/m³ breathable air. Assume 1.2 m³/h respiration rate, 70 kg body weight, and 75% retained and 100% absorbed.

Note: Here intake (I) is marked as I_C to denote "carcinogenic."

Solution: IRIS data shows nickel dust as a class A carcinogen (inhalation SF/CPF of 8.40E−01)

$$I_c = \frac{[C] \times (CR) \times (EF) \times (ED) \times (ABS) \times (RR)}{(BW) \times (AT)}$$

$$= \frac{\left[0.5\,mg/m^3\right] \times \left(1.2\,m^3/h\right) \times \left(8\,h/d\right) \times \left(230\,d/y\right) \times \left(8\,y\right) \times \left(1.0\right) \times \left(0.75\right)}{\left(70\,kg\right) \times \left(25{,}550\,d\right)}$$

$$\boxed{I_c = 3.704 \times 10^{-3}\ mg/kg \cdot day}$$

Notes: (CR) the respired air is usually given per hour. We need **daily** rate. If intake was for a 24 h/d resident, the /h rate would be multiplied by 24 h/d, but the worker is only present 8 h/d, so we could enter the CR values as: $(1.2\ m^3/h) \times (24\ h/d) \times (8\ h/24\ h)$

but this can be simplified to: $(1.2\ m^3/h) \times (8\ h/d)$

Intake Calculation 6: Unknown Airborne Carcinogen in School

Determine the chronic daily inhalation intake in children aged 6–12 of an <u>unnamed</u> carcinogenic material. Assume continuous (residential) exposure (all day, every day) for 2 years.

In this instance, we leave the concentration as [C mg/m^3]. We will be calculating only the intake of AIR/kg $\cdot$ day

$$I_c = \frac{[C] \times CR \times EF \times ED \times RR \times ABS)}{(BW \times AT)}$$

Assumptions and values used:

CR = 0.46 m^3/hr × 24 h/d = 11.04 m^3/day …. (Fig. 3.20)

BW = 29 kg …. (Fig. 3.20)

EF = 365 d/yr. …. (given)

ED = 2 yr. …. (given)

RR = ABS = 1.0 (100%) as in the previous example

AT = (365 × 70) days (for a carcinogen, this must be the lifetime of days)

$$I_c = \frac{[C] \times (11.04\,m^3\,/\,day) \times (365\,d\,/\,yr) \times (2\,y) \times (1.0) \times (1.0)}{(29\,kg) \times (25550\,d)}$$

The CDI can be written as:

$I_c = (0.01088 \times 10^{-2}\ m^3/kg\cdot day) \times [C\ mg/m^3]$

Note that when the air intake (in $m^3/kg\cdot d$) is multiplied by the missing concentration (in mg/m^3), the product is correct as $mg/kg\cdot d$.

Example #6 calculated the intake of the medium (in this case air), leaving the concentration [C] incomplete. This is a preferred way to complete an analysis where people are being exposed to several substances that are all carcinogenic (or noncarcinogenic) and all via the same media/pathway. The intake x [C] value will be the same for all these substances, with just the concentrations remaining to be factored in. An example is shown in Fig. 3.23.

Intake Calculation 7: Unknown Noncarcinogenic Solvent in Workplace
Calculate the intake for a 70 kg adult's 5-year occupational exposure to an airborne noncarcinogenic solvent. Assume 8-hr working day for 200 working days/yr, light work (a respiration rate of 1.2 m^3/h), 100% absorption, and 70% retention.

We can calculate [C] x intake as follows:

$$I_N = \frac{[C] \times CR \times EF \times ED \times ABS \times RR}{BW \times AT}$$

$$I_N = \frac{[C\,mg\,/\,m^3] \times 1.2\,m^3\,/\,h \times 8\,h\,/\,d \times 200\,d\,/\,y \times 5\,y \times 1.0 \times 0.70}{70\,kg \times (5 \times 365)\,d} =$$

$I_N = [C\ mg/m^3] \times 0.052603\ m^3/kg\cdot d$ **The concentration can be entered when available.**

3.5   Risk Characterization [Step 4]

3.5.1   Translating Chronic Daily Intake to Risk

The contaminants have reached the human being through various pathways, and the quantities inhaled, ingested, absorbed, and retained have been estimated. The next step is to predict the probability of adverse health effects and interpret the implications. We begin with the chronic daily intake (I or CDI), or LADD (lifetime average daily dose if the assessment is for a lifetime exposure), and then use the most current indicators of the potential for this substance to cause cancer (the slope factor, SF), or the potential to cause noncarcinogenic illness (the "reference dose," RfD) to "characterize" the risk.

3.5.2   Lack of Epidemiological Data

We are predicting the risk potential in quantitative terms about human health to the community, society, or to specific populations, such that they can make informed decisions about their health and safety. Clearly, then, we should be basing the assessment on solid epidemiological data from human studies, but these data are rarely available. Experimenting on human subjects in modern times is restricted by ethical barriers, and limited to low-risk trials or observational studies. Studies of human exposure to dangerous substances are usually limited to "observational" (non-experimental) studies of workers' exposures in the workplace, or of infrequent accidental exposures, spills, releases, or other "natural experiments." In neither case are the data reliable or valid enough for precise reference doses.

Other reasons that human data are not commonly available include the typical latency period in humans before any effects (adverse or otherwise) can be expected. Unlike small animals, this can be relatively long. In small rodents, the expected life span may be 1.5–2.5 years, while birth to reproductive maturity in mice is 5–6 weeks, compared to 16–20 years in humans. Also, humans are often exposed to complex and non-recorded mixtures of chemical, biological, and physical agents, unlike lab animals, which are also bred to reduce extraneous genetic factors and other potential confounders.

For this reason, in vitro (test tube) and in vivo (animal) studies are the source of most of the slope factors (SF) and reference doses (RfD) used in risk assessment. This of course requires allowances for animal-human extrapolation.

3.5.3 Do We Use the "Mean" or "Maximum" Concentration Data?

Either mean or maximum concentrations can be used to estimate risk provided you clearly explain which you have used. However, performing a specific risk calculation using *both* values permits the estimation of a range of potential risks. The use of mean values represents a more realistic estimate of long-term, chronic exposure scenarios for most human subjects. Maximum values are best used in shorter-term, sub-chronic risks, and also to provide a useful upper-bound estimate of potential risk. Stochastic methods to assess concentrations using multiple input measurements are discussed in Sect. 3.7.

Risk assessment guidance has tended to emphasize the use of a single estimate of exposure concentration in the calculation of potential risks. This value, described as the "upper bound," is often the 95 percent upper confidence limit around the data mean (Fig. 3.13). Use of this number is meant to provide an estimate of risk that is already worse than most (97.5%) of the population. But arguments can be made that this contributes to a worrying overestimation of potential risk, especially when used in combination with other worst-case assumptions to define an exposure. A more focused discussion of the overestimation of risk is found in Sect. 3.6.

3.5.4   Calculation of the Carcinogenic Risk

Carcinogenic risk is estimated as the chronic daily intake, or CDI (the intake or I_C, calculated in the exposure assessment stage), multiplied by the "slope factor" (SF) or "carcinogenic potency factor" (CPF). The product is the probability (without units) of excess lifetime cancer from exposure to this chemical under the included assumptions, conditions, and parameters. Put more simply, it is the additional (or incremental) risk of developing cancer due to the specified exposure. The calculation is:

$$\textbf{Carcinogenic Risk} = \textbf{CDI} \times \textbf{SF}$$

Note in Fig. 3.21, the units cancel completely, leaving a unitless probability value of 2.60×10^{-5}. *What exactly does this value represent?* An average individual (with the age, weight, etc., we considered in the calculations), who drinks 2 L of this water every day for a lifetime, has an expected *additional* or *incremental* possibility of death from cancer of 2.6/100,000, or 26 chances in a million, in addition to the approximately 230,000 chances in a million (background risk) that they already carry just by being human in an industrialized country. For the population as a whole, it implies 26 additional deaths per million population over a lifetime.

Although we have different ways to express the calculated risk, we still need a *judgment* as to this figure: Does it represent a serious threat or is it too trivial to worry about? Does it exceed a legislated or regulated standard or limit?

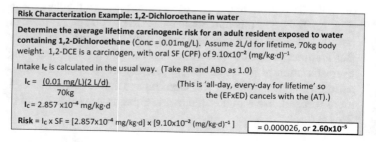

Fig. 3.21 Risk characterization: (1,2-dichloroethane in water)

3.5.5 Use of *de minimis* to Characterize Carcinogenic Risk

Strict limits or regulated standards do not exist for carcinogenic risks.[5] Instead, in characterizing an assessed risk, a judgment is made, based loosely upon what society has determined to be an "actionable level" of carcinogenic risk above the background risk. For the general public, this is the "de minimis"[6] level of 1 in a million (1.0×10^{-6}). The level of "acceptable" risk may be a different when applied to the workplace.

Above that level, we may expect public outcry for action. But while the de minimis of one-in-a-million is commonly cited in decision-making and goal setting, we cannot assume that society, or sections of the community, will always see the (1×10^{-6}) value as a cutoff for concern; they may see much smaller risks as unacceptable while voluntarily accepting and ignoring risks thousands of times greater. Although it has no scientific or legislative basis, 1.0×10^{-6} is the point above which society is expected to become uncomfortable and concerned enough to demand that *something should be done* to rectify the situation.

From the dichloroethane example in Fig. 3.21, the incremental carcinogenic risk of 0.000026 is 26 times the de minimis level of 1×10^{-6} (0.000001), and as such, the risk to an individual consuming this water for a lifetime can be said to be around 26 times higher than what society would expect and accept.

It is important to remember that this risk is only from the water. It is the incremental risk due to consuming *that* water over a lifetime, and says nothing about other exposures, or the background risk (which is invariably much higher).

---

[5]Early attempts to prohibit substances that demonstrated *any* carcinogenic potential were quickly found to be unworkable as analytical methods became more powerful (see Delaney Clause below).

[6]**De minimis:** "Level of risk that can be ignored." From common law maxim: *de minimis non curat lex* (the law does not concern itself with trifles).

Put another way, if a million people consumed 2 L/d of this water for their entire lifetime, 26 additional deaths would be expected to result.

3.5.6   The Delaney Clause

The Delaney Clause is a 1958 amendment to the US Food and Drug Act of that forbids the addition to food of any additives shown to be carcinogenic in any species of animal or in humans. As each decade passed, technology became increasingly able to detect minute amounts of almost any substance in any food, and the provisions of the Delaney Clause has for some decades been considered well intentioned but unrealistic (Ames 1990; Swirsky-Gold et al. 1987). Delaney is still on the books and is sometimes cited during protests against new additives or ingredients.

3.5.7   Options for Reducing Carcinogenic Risk

Whether from food, water, or breathable air, a risk that exceeds the socially acceptable value of 1 in a million presents three hypothetical remedies for reducing the risk (not all of them are practical options in each circumstance):

1. Reduce the contact rate (exposure to the food, water, air or other medium)
2. Reduce the concentration of the contaminant in the medium
3. Change to a safer source

The analysis in Fig. 3.21 involves a contact rate for water, at 2 liters a day, and to reduce the risk such as not to exceed de minimis, these options translate to (1) drinking only 1/26th of 2 L (= 77 ml/day) of this water, or (2) treating the water such that the concentration of 1,2-DCE is reduced to 1/26th of 0.01 mg/L (= 0.385 µg/L) or (iii) changing the source of water. Clearly, in this example, drinking only 77 ml/d of water (1) is unrealistic.

Removing most of the contaminant (2) might be attainable by some form of treatment either at source or at point-of-use while selecting an alternative source (3) presents a more immediate resolution.

3.5.8 Calculation of a Noncarcinogenic Hazard Index

The equivalent of a risk determination for a noncarcinogenic exposure is the *hazard index (HI)*. Both the calculation and the interpretation are different. The intake (I_N) is divided by the RfD:

$$HI = I_N / RfD,$$

As can be seen in Fig. 3.22, the units again cancel out, leaving the unitless probability known as hazard index (HI). Again, no moral or legislative standard exists for limiting toxicants, but an HI that exceeds 1.0 can be considered excessive in general settings. In some workplace settings, occupational exposures have carried an elevated risk that was assumed to be part of the work,

---

Risk characterization example: (1,2-Dobromo-3-chloropropane in air)

Calculate the inhalation hazard Index (HI) for exposure in the workplace to 1,2-Dibromo-3-chloropropane (DBCP) at mean level of 0.004 mg/m^3.

Assumptions: 8 hr shifts; 200 working days a year for 10 years; 70 kg body weight; respiration rate 0.90 m^3/h.   EPA-IRIS shows DBCP has inhalation RfD of 5.72x10^{-5} mg/kg·d

I_N = $\dfrac{[0.004 \text{ mg/m}^3] (0.90 \text{ m}^3/\text{h}) (8\text{h/d}) (200\text{d/y}) (10\text{y})}{(70\text{kg}) (365\text{d/y}) (10\text{y})}$ = 0.00022544 mg/kg·d

HI = I_N/ RfD, = (2.254x10^{-4} mg/kg·d) / (5.72x10^{-5} mg/kg·d) | **= 3.94**

Using RfC method rather than conversion to inhaled dose: Calculate the inhalation hazard Index (HI) for exposure in the workplace to 1,2-Dibromo-3-chloropropane (DBCP) at mean level of 0.004 mg/m^3. The RfC of DBCP is 2x10^{-4} mg/m^3. (See section 3.5.9 for discussion)

HI = [C] / RfC,
HI = (0.004 mg/m^3) / (2x10^{-4} mg/m^3) = 20

Fig. 3.22 Risk characterization: (1,2-dibromo-3-chloropropane in air)

with those involved being fully aware of the risk. Hazard indices as high as 10 have been deemed acceptable, but this decision should occur on a case-by-case basis.

In the example (Fig. 3.22), the assessed HI value based on the RfD exceeds 1.0 by nearly four times, so we can assume the community of workers would expect some action to reduce the risk.

Risk reduction Options Three theoretical (but not always practical) options for bringing the HI to 1.0 or less in this setting would include (1) reducing the contact rate (CR), for example, breathing only ¼ as much air while at the workplace (clearly not practical); (2) reducing the concentration of DBCP in air to about one quarter (i.e., 0.001 mg/m³ of air) by source control, filtration, ventilation, extraction, etc.; or (3) by breathing air from a different source (e.g., by using self-contained breathing apparatus, or providing ducted ventilation to the area, or by moving to another location altogether), certainly in the short term, a realistic solution.

3.5.9   RfC and IUR Use

In the majority of the work described herein, we chose to convert the external inhaled exposure into an inhaled dose by evaluating breathing rates, contaminant concentrations, etc. However, it is also possible to calculate a quick hazard index by simply comparing the contaminant concentrations to their published reference concentrations values. For hazard index, this is accomplished by dividing the concentration by the RfC value.[7] For carcinogenic compounds, the IUR can be multiplied by the estimated lifetime exposure to calculate cancer risk. The advantage to using an inhaled dose rather than the external air concentration with accompanying RfC or IUR is that it can account for specific exposure scenarios that may alter uptake of the contaminant.

---

[7]This "simplified" process somewhat reflects the toxicity score demonstrated in Sect. 3.2.5.

3.5.10 Characterizing Groups of Carcinogenic Substances in the Same Medium

Consider a community's 10-year exposure to a group of eight waterborne chemicals in their rural water supply. The intake of the medium (water) will be the same for all of these substances, allowing the I_C for each carcinogen to be completed but awaiting the concentration value:

$$I_C = \frac{[C\,mg\,/\,L] \times (2\,L\,/\,d) \times (365\,d\,/\,y) \times (10\,y)}{(70\,Kg) \times (25,550\,d)}$$
$$= [C\,mg\,/\,L] \times (4.08 \times 10^{-3}\,L\,/\,kg \cdot d)$$

and this value can be placed at the head of the "intake" column in Fig. 3.23:

The intake (I_C) for each chemical can now be computed by multiplying the individual concentration [C] by 0.00408 to complete the water intake value (I_C).

For instance, for di(2-ethylhexyl) phthalate (DEHP), a plasticizer:

$$I_C = (0.50000\,mg\,/\,L) \times (4.08 \times 10^{-3}\,L\,/\,kg \cdot d) = 0.0020400\,mg\,/\,kg \cdot d$$

	[C] max mg/L	oral SF(CPF) 1/mg/kg·d	INTAKE (Ic) [C] x 4.08×10⁻³ (mg/kg·d)	RISK Ic × SF
DEHP	0.50000	0.01400	0.0020400	0.0000286
Heptachlor	0.00045	4.50000	0.0000018	0.0000081
Pentachlorophenol	0.00844	0.12000	0.0000344	0.0000041
Formaldehyde	0.00453	0.04500	0.0000185	0.0000008
1,2-Dichloroethane	0.00061	0.09100	0.0000025	0.0000002
Chloroform	0.00052	0.00610	0.0000021	0.0000000
Mirex	0.00005	no data	-	-
Chlorobenzyl	0.00031	no data	-	-
				0.00004186

Fig. 3.23 Risk characterization: groups of carcinogens (rank-ordered by risk)

The risk is then obtained by multiplying the intake by the SF(CPF):

$$\text{DEHP risk} = \left(0.0020400\,\text{mg}/\text{kg}\cdot\text{d}\right)\times\left(0.014001/\text{mg}/\text{kg}\cdot\text{d}\right)$$
$$= 0.0000286$$

The risk for each chemical can now be characterized separately to identify substances associated with a particularly high risk that might be controllable through treatment. We can also assess the risk presented by the whole group as well (4.186×10^{-5}).

In this example, DEHP is already >28 times the de minimis level, while heptachlor and pentachlorophenol exceed 1×10^{-6} by 8 and 4 times, respectively. The remaining two carcinogens, when considered separately, are far below de minimis together and, even when considered together, do not exceed 1×10^{-6}. These latter two might be acceptable if the top three components were removed.

3.5.11 Characterizing Groups of Noncarcinogens in the Same Medium

The noncarcinogenic exposure table for the same eight chemicals is shown in Fig. 3.24. The water intake (I_N) was calculated (omitting the concentration) as:

	[C] max mg/L	oral RfD mg/kg·d	INTAKE (In) [C]x 0.0286 (mg/kg·d)	Haz Index (I_N/RfD)
Mirex	0.00005	0.000002	0.00000143	0.71500000
Heptachlor	0.00045	0.000500	0.00001287	0.02574000
Pentachlorophenol	0.00844	0.030000	0.00024138	0.00804613
Chloroform	0.00052	0.010000	0.00001487	0.00148720
Formaldehyde	0.00453	0.200000	0.00012956	0.00064779
Chlorobenzyl	0.00031	0.020000	0.00000887	0.00044330
DEHP	0.50000	11500.00	0.01430000	0.00000124
1,2 Dichloroethane	0.00061	no data	-	-
				0.75136566

Fig. 3.24 Risk characterization: groups of non-carcinogens (rank-ordered by HI)

$$I_N = \frac{[C \, mg / L] \times (2 \, L / d) \times (365 \, d / y) \times (10 y)}{(70 \, Kg) \times (10 \times 365 \, d)}$$
$$= [C \, mg / L] \times 0.0286 \, L / Kg \cdot d$$

and this value can be placed at the head of the intake column.

The risk equivalent for noncarcinogenic agents is the hazard index (HI), obtained by *dividing* the intake (I_N) by the RfD. For example,

For instance, for Mirex (a banned organochloride insecticide):

$$I_N = (0.00005 \, mg / L) \times (0.0286 \, L / kg \cdot d) = 1.43 \times 10^{-6} \, mg / kg \cdot d$$

The risk is then obtained by dividing the intake by the RfD:

$$HI(Mirex) = (1.43 \times 10^{-6} \, mg / kg \cdot d) / 2 \times 10^{-6} = 0.715000$$

The overall hazard index for this example is 0.715, which is less than the level expected to promote concern, usually understood as 1.00. Again, the HI is not a legislative or statutory standard, but by convention, the level below which we assume society to be willing to accept the exposure. The consumer's "risk" from water exposure under these parameters does not seem excessive and should not induce community unrest. If the HI *had* exceeded 1.00, we could suggest that reducing the original concentration of the chemicals (either targeting one or all) would allow the HI to not exceed 1.00. This would be accomplished by dividing the original concentration by the HI.

3.5.12 Conversions: ppm to mg/L etc

Not all testing laboratories report concentrations in the same way, and the literature will often report parts per million (ppm) or parts per billion (ppb), whereas risk assessment procedures expect mg/L for water, mg/m^3 for air, or mg/kg (food).

Foods and Other Solids: Measured in kg, the conversion is simple. Milligrams (mg) are already a measure of mass, so 1 ppm = 1 mg/kg, and 1 ppb = 1 μg/kg.

Liquids: 1 ml water weighs 1 g (1,000 mg); therefore, 1 mg of a liquid with the same density as water in 1 L of water is 1 ppm. Any solid dissolved in water probably has a different density, meaning that a conversion is needed for each substance to be accurate. However, for practical applications with dilute aqueous solutions, 1 mg/L can be considered to be close to 1 part per million (ppm). For example, a sodium nitrite concentration of 1 mg/L can be taken as 1 ppm nitrite.

Air: Vapors and gases have different densities, so a conversion is required. You will need to find or calculate the molecular weight of the gas. Conversion calculators are available online, for example, NIOSH at http://niosh.dnacih.com/nioshdbs/calc.htm.

3.6   Critical Perspectives of the Risk Assessment Process

3.6.1   Excessive Use of "Worst-Case" Compounded

At every juncture, for every variable and decision-making point, the highest level of risk is typically employed, beginning with the still common use of the *"maximum"* concentration (usually the upper 95% "bound" or confidence limit), from the values found on site in soil, food, water, or air, instead of the observed *"mean"* level.

3.6.2   The Toxicological Parameters

The toxicological parameters used to evaluate and estimate the potential degree of harm from an exposure are derived from established points on the dose-response curve (e.g., LOAEL), but then adjusted for LOAEL-to-NOAEL extrapolation, animal-to-human extrapolation, chronic-to-sub-chronic extrapolation, and so on.

The adjusted value for a reference dose (RfD) may be 2–4 orders of magnitude lower than the lowest adverse effect actually observed (Fig. 3.15).

3.6.3 "Conservative Estimates"

Further confusing the issue from the perspective of the public or the media is the use of the term "conservative." In everyday usage, this implies a value that is slightly *lower* than expected, or *less* than the likely mean average derived from the data. In the field of risk assessment, however, a "conservative" estimate is intended to convey a value that has been *increased*, often to the upper 95 percent confidence limit above the mean. The reason, once understood, is clear: if we assess, characterize, and act on the risk for a *maximally exposed individual* (MEI), we will have protected the overwhelming majority of the rest of the population. But does credibility suffer when we assume that the individual will consume 2 liters of *this* imported bottled water every day for their lifetime? The hypothetical *"maximally exposed individual"* (MEI) lives an entire lifetime at the point where pollutant concentrations are highest, or regularly consumes the highest-contaminated food and water. The *most likely cancer risk* from power plants in the USA, for example, was estimated by Vallero (2010) to be 100- to 1000-fold lower for an average exposed individual than that calculated for the MEI.

Keenan, writing in the journal risk analysis more than 25 years ago, had already noted the increasing tendency to overstate the risk by compounding the worst-case scenario. While the US EPA's maximally exposed individual (MEI) is a useful tool, it must be remembered that the relevance is limited when applied to *human risks that are likely to be encountered*. Multiple estimates of risk, stratified by exposure, would be a useful option.

A splendid example of repeated use of "conservative" assumptions is provided by Keenan (1994) in the example of an assessment of dioxin hazard from pulp and paper mill sludge applied to crop-growing lands. The estimate of lifetime cancer risk for members of a subsistence family had been made, assuming each member of the family would consume more than 8.2 lb (>3.7 kg) per day of beef, pork, chicken, and dairy products produced on the farm, together with catfish fed with contaminated soybean over a 70-year lifetime. The origin of the dioxin in the food had been based on sludge application to fields that were more than 25 times the amount actually applied, fish bioaccumulation factors more than 30 times higher than the maximum values found in the literature, and sludge dioxin levels 40 times greater than the level indicated in the data. The official carcinogenic risk was in the vicinity of 10^{-2}. But taking the "reasonable daily intake" of meats/fish to be 2 lb ($\approx$1 kg) per day (or ¼ of the estimate used), and similarly factoring in "reasonable" levels of the other values, the official estimate may be as much as 100,000 times higher than an estimate based on more realistic or reasonable values.

The use of more informative parameters (such as arithmetic or geometric means, or medians), and stochastic (e.g., *Monte Carlo*) techniques would also provide more realistic ranges/*distributions* of probabilities instead of maximal deterministic estimates (see Sect. 3.7).

3.6.4   Missed Transformation and Transport Mechanisms

The environmental fate of pollutants and environmental toxicants may not be fully understood or even imagined during the early stages of an assessment. Many biological, biochemical, and physical factors are active in soil, humus, and exposure to sunlight and weather elements. Keenen (1994) cites an example of the assessment of dioxin vapors from an incinerator. The initial concern for exposure was eliminated when it was learned that dioxin as a vapor has a half-life of about 90 minutes. In contrast, 2,3,7,8-tetrachlorodibenzo-p-dioxin (TCDD) in soil and fly ash may have an environmental half-life of 12–50 years.

Similarly, indirect pathways may escape consideration such as settlement on pasture and livestock to appear later in dairy products, eggs, and meat.

3.6.5 Inappropriate Statistical Assumptions and Methods

Environmental and occupational data do not commonly demonstrate a "normal" distribution. Log-normal distributions, for example, are more frequently found. When guidelines call for standard deviations or the 95% confidence limits, the result will be greatly overestimated if the distribution is lognormal and an appropriate data transform has not been made.

Similarly, where no detectable level of a contaminant is found in a majority of samples at a site, analysts have in the past used the minimum detection limit (MDL) or one-half of the detection limit, on the premise that the material may be present at that level. Travis and Hester (1990) have suggested an approach to avoid this overestimation.

3.6.6 Overuse of Anthropometric Parameter Tables

Tables such as Fig. 3.20 are valuable sources of anthropometric and exposure data, but if actual measurements are available, they should be used. Population norms expressed, for example, as BMI and other parameters have changed over the last few decades and between regions. Depending upon their geographic and ethnic origins, adults are certainly not necessarily at a mean 70 kg in weight. Gender-specific differences in exposures, metabolism, and adverse effects have also been generally ignored in observational studies and even randomized trials.

3.6.7 Questionable "Additive" Model for Noncarcinogens

Some authorities label the noncarcinogenic risk for individual substances the *"hazard quotient"* (HQ), with the *"hazard index"* (HI) being the total of all substances present in that medium or that exposure (HI = $\sum$HQ). Regardless of the nomenclature, the standard additive model whereby the hazard index is the "composite" for all substances under study presents a difficult interpretation. Each toxicant usually targets a certain organ system or set of tissues, and adding together measures of adverse effects from several different targeted systems makes little sense physiologically.

As an example, if an exposure resulted in a *hazard index* of 0.5 for diethylene glycol, and 0.4 for methanol, and 0.3 for paraquat, the total HI would be 1.2, suggesting an unacceptable adverse health effect requiring remediation. However, since DEG's metabolic by-product, oxalic acid, targets renal tubules, while paraquat's acute effects are upon the lungs, and methanol damages the optic nerve, it is unreasonable to accept that the sum of the three HIs is a true indication of the adverse burden upon the body.

One approach could be to group noncarcinogens by the organ systems targeted and develop an HI for each group, for example, all toxicants having an effect upon the lungs or nervous system.

3.6.8 Omitted Relationship to Background Risk

An exposure that by itself is calculated to present a certain adverse effect or incremental risk can become the focus of public and media concern to the exclusion of the background risk for the same outcome. The background risk, calculated across the population as a whole, in many instances is vastly greater than the incremental risk being studied and of course will remain almost unchanged regardless of any remedial action for the present challenge. For objective and constructive planning, assessment, and remediation, the incremental and background risks should be compared and contrasted.

This is an especially important consideration when we are reminded of the considerable difficulties the public already has in understanding the process of risk assessment. For an illustration of the benefit of applying this concept in the realm of risk communication, please refer to case study #14 at the end of Chap. 6.

3.6.9 The Need to Consider Variability and Uncertainty

The risk assessment process is, by its predictive nature, accompanied by uncertainty at every juncture. Numerical and probabilistic safety margins and adjustments are included in the estimates, but it is important to accompany any estimate of risk by a discussion of possible inaccuracies, interpretations, and especially the full list of assumptions deployed in arriving at the final risk or hazard indices.

Uncertainty and variability are from distinct sources and should be reported and dealt with separately. True **variability** arises from the characteristics of that variable in its "natural state," meaning that if we are describing a sample of people from a population, each member of that target population, from which the sample is drawn, carries a biological variability that is normal and expected. It cannot be reduced or eliminated because to do so would be to change the criteria for selection, and the sample would no longer represent the population; we would have biased the selection and made spurious any results. We cannot change biological variability, but we can measure it and ensure it is accounted for in the analysis by means of confidence intervals.

True **uncertainty** can arise from a lack of information, insufficiently sized dataset, a sample size that is too small to achieve a valid measurement, or from missing data.

In addition, the assessor needs to be fully aware of **perceived uncertainty** of risk, hazard, and danger that can be generated as a result of reading the document or seeing/hearing discussions about it. For example, the mainstream media, in attempting to offer fair and equal voicing, may feature a discussion or interview with a representative from both sides of the issue, suggesting, ostensibly, to the listening or watching public, a 50–50% representation. The actual consensus of global experts, however, may be 99.2% vs. 0.8%, but this attempt at media "fairness" can actually reinforce support for the minority, who have now been given the idea that the two schools of thought are about equally divided.

Another source of uncertainty leading to potential misperception and inadvertent "misdirection" is the use of the phrase "conservative estimate." In the language and meaning of risk assessment, unlike in everyday usage, a "conservative estimate" is an *overstate-*

ment of the risk and the factors leading to the estimate. This is discussed further in Sect. 3.6.3. Someone not understanding this concept could be forgiven for assuming that if this is a "conservative estimate" of the risk, then it probably is much greater. Bringing this to light is good advice whenever the opportunity permits.

3.7   Deterministic Versus Stochastic Risk Assessment

Up to this stage, the calculations have resulted in *deterministic* estimates of risk, that is, data values expressed as a single numerical value, a best estimate within the confidence intervals and with as much precision as the predictive methods would allow. Many risk assessments still use this model, with single-point values entered for each variable in the intake calculation: body weight, exposure frequency, duration, contact rate, and so on. Point estimates, however, are inflexible and provide no information as to the distribution, range, and characteristics of the variable. Are all members of the subject population close to 70 kg in weight, for example? Or are they inclusive of a wide range of phenotypes and ethnicities?[8] Deterministic techniques also provide very little information about uncertainty and variability surrounding the final risk estimate and is even less able to distinguish between the two (Sect. 3.6.9. discusses this in detail) (EPA (US)a 2022).

These deficiencies can be partly addressed by using a **probability distribution** in place of each point estimate, in a process called **stochastic modelling**.

Stochastic modelling treats variables as distributions of data and selects values at random from within each distribution. Each variable will have its own characteristics, and a first step is to identify the shape of each distribution.

For instance, the body weight for the studied population may be available as a log-normal or Weibull distribution, whereas the contact rate (in L/day) may best be identified as histogram

---

[8] It has often been observed that the "average" human being has one testis and one ovary, although this is far from a common finding.

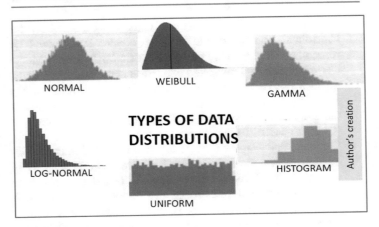

Fig. 3.25 Examples of data distributions

(Fig. 3.25). Commercial software is available to help identify and "best-fit" each distribution from the available data.

Other software components then calculate the intake (I) from perhaps 100,000 or half a million iterative calculations, each time taking a separate value, randomly sampled from within every distribution. At the end of the process, the final result can be a range of probabilities for each stratum, or percentile, quartile, quintile, etc. The software accomplishes this task by running "underneath" a spreadsheet system such as Microsoft's® Excel®. The terms *"Monte Carlo"* and *"Latin hypercube"* are two such randomization procedures that sample from each distribution using a different probability algorithm.

Some published examples of stochastic applications follow:

1. Thompson et al. (1992) assessed children's exposure to a carcinogenic contaminant in soil through ingestion. Of 17 parameters influencing the intake of the carcinogen, 12 were entered as distributions, rather than single point values.
2. Evans et al. (1992) examined carcinogenicity risk from formaldehyde but reported that estimates of model uncertainty between 50th and 90th percentiles were extremely large.

3. Differences in Lifetime Excess Cancer Risk (LECR) for Canadians from food and beverages were investigated by researchers at the University of Victoria (BC). Using Monte Carlo techniques, they found two substances (lead and PERC) to have excess risk below 10 per million, whereas for the remaining three (arsenic, benzene, and PCBs), at least 50% of the population were above 10 per million excess cancers (Cheasley et al. 2017).

4. The number of ingested bacterial cells expected to cause salmonellosis was investigated by Bollaerts et al. (2008). They modelled the dose-illness relationship based on data from 20 Salmonella outbreaks using two "bootstrap" methods. A first procedure accounted for stochastic variability whereas a second procedure accounted for both stochastic variability and data uncertainty. The analyses indicated that the more susceptible population has a higher risk (probability) of illness when the dose is low if the pathogen-food combination matrix is extremely virulent, while the risk is high at high-dose levels when the combination is less virulent.

5. Early in the Covid-19 pandemic, transmission of the SARS-CoV-2 virus in Australia was predicted through modelling by Xie (2020) using a Monte Carlo technique. The eight key parameters and assumptions were:

- Observation period (in days) of a simulation study
- Simulation period (in days) of a simulation study
- Average reproduction/infection rate (i.e., the expected number of secondary cases that each existing infectious case will generate: R_t)
- Average/expected number of days for an existing infectious person to infect a susceptible person in the population mean parameter for the negative binomial distribution
- The dispersion parameter for the negative binomial distribution
- The study/target population size
- The proportion of people with immunity in the population
- The initial number of infectious persons prior to the observation/simulation period (Fig. 3.26)

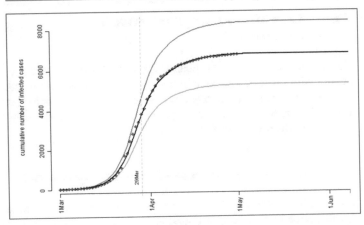

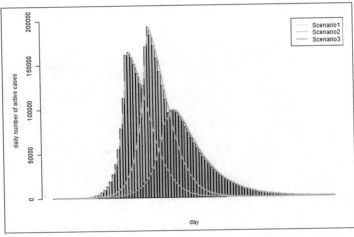

Fig. 3.26 (**a**) The accumulative number of confirmed cases over the observation period: the bold curve is the model estimated median level, the upper curve is the 25th percentile level, and the lower curve is the 75th percentile level out of 1000 bootstrap simulation runs; the dot points are the actual recorded number of cases from the UK. (Xie 2020) (Reprinted from Xie 2020, Fig. 4. https://doi.org/10.1038/s41598-020-70091-1, licensed under the terms of the Creative Commons Attribution 4.0 International License (http://creativecommons.org/licenses/by/4.0/)). (**b**) Predictions for the growth of the pandemic could be modelled, giving three hypothetical infection rate patterns and predicting the impact upon population and immunity level (Xie 2020) (Reprinted from Xie 2020, Fig. 7. https://doi.org/10.1038/s41598-020-70091-1, licensed under the terms of the Creative Commons Attribution 4.0 International License (http://creativecommons.org/licenses/by/4.0/))

3.8   Case Studies in Quantitative Risk Assessment

These nine case studies in quantitative risk assessment are presented with most of the calculations or interpretation completed, but there will be some missing components for you to supply. Try each of these and then compare your conclusions with the solution table following case #11.

3.8.1   Case Study #3: US EPA Recall Due to Arsenic

On March 7, 2007, the EPA issued a warning to consumers and a recall notice to suppliers, vendors, and wholesalers in respect of imported bottled water from an east European country (ProMED 2007).

Case Study #3: US EPA Recall of Water Containing Excessive Arsenic (2007)
"The US Food and Drug Administration (FDA) is warning consumers not to drink certain brands of imported mineral water due to arsenic content ...."
Product: 500 mL green glass bottles of mineral water containing 500–600 micrograms (µg) of arsenic per liter. Labels show more than 30 brand names, but all originate from the same producer. FDA's standard for bottled water allows no more than 10 micrograms per liter. The recall required removal of this product from the marketplace and distribution. The manufacturer and the importer argue that this is unreasonable.
Calculation: Assume consumption of this water at 2 L/d for life (70 years), an arsenic concentration of 500 µg/L (0.5 mg/L) and adult body weight of 70 kg. EPA-IRIS RfD for inorganic arsenic is 3.0×10^{-4}. Calculate the non-carcinogenic lifetime average daily dose (LADD) of arsenic by this pathway: $$\text{Intake}\left(I_N\right) = \frac{[C] \times (CR)}{BW} = \frac{(0.5\,\text{mg}/\text{L}) \times (2\,\text{L}/\text{d})}{70\,\text{Kg}}$$ $$= 0.01429\ \text{mg}/\text{kg} \cdot \text{d}$$

$$\text{Haz.Index}\left(\text{HI}\right) = \frac{I_N}{\text{RfD}} = \frac{0.01429}{0.0003} = 47.63$$

Challenge: [A] Given the calculations shown here, was the EPA recall valid? [B] A reporter asks you to explain how much of this water could a person drink in a day on a permanent basis and still remain within the EPA's safe standard. [C] How would you assess the EPA's recommended standard. It is adequate? Excessive?

3.8.2   Case Study #4: Hazardous Waste Site, Lackawanna, Pennsylvania

This was a disused open-pit coal mine that was licensed in the 1970s for receiving municipal and commercial refuse and non-hazardous sludge. But later that year, it was discovered that hazardous/toxic waste and industrial waste were being disposed at the site. The permits were rescinded.

Case Study #4: Hazardous Waste Site: Lackawanna, Pennsylvania (1978–85)
The Lackawanna site (Pennsylvania) site was a former open-pit coal mine licensed as a landfill to receive only municipal waste. It was later discovered being used illegally as a dump for drums of hazardous waste. Most of the drums had been crushed when placed in the landfill, but some had remained intact. The existence of intact drums of unknown contents prevented normal exploratory drilling, and special procedures were required to investigate subsurface conditions (EPA(US)c 2019).
The EPA Record of Decision (RoD) in March 1985 [Excerpt]   *"…operators of the trucking firm testified that they had brought drums of hazardous waste to the site … Estimates ranged between 10,000 and 20,000 drums…"*
Hazard Identification: More than 80 chemicals were found at the site. They were grouped into 13 "surrogate" chemical types and 6 metals:

Arsenic	Trichloroethylene	Methylethyl ketone	Copper
Acetone	4-methyl-2-pentanone	Toluene	Lead
Cresol	Tetrachloroethylene	Benzene	Mercury
2-hexanone	1,2-dichloroethane	[Metals]	Cadmium
Ethylbenzene	Methylene chloride	Barium	Chromium
xylene			

Exposure assessment: The migration of each indicator compound to the nearest receptors was modelled, including the use of the mine pool under the site as a municipal water storage.

Toxicity assessment: Slope factors (SF) for carcinogens and reference doses (RfDs) for noncarcinogens were identified. Where no values were available, toxicologists used the latest research information and proposed EPA's Minimum Concentration Limit Guidelines (MCLG) for drinking water.

Risk characterization: A "No action" option, and three alternative options were investigated:

Options	Carcinogenic risk
A: (No action)	0.55×10^{-6}
B: Cap and leachate treatment	0.44×10^{-6}
C: B and drum removal	0.23×10^{-6}
D: C and removal of contaminants	0.14×10^{-6}

Challenge: Provide the "characterization." Note: The EPA was using an "acceptable" risk range of 10^{-4} to 10^{-7} at the time (1985), but when compared with the current 1×10^{-6}, do any/all of the options – including the "No action" option, meet EPA's level of acceptable risk?

3.8.3   Case Study #5: Poyang Lake (鄱阳湖)

The largest freshwater lake in China plays an important role in regulating the Yangtse River level, conserving water resources, and maintaining the ecology of surrounding areas. In a well-designed study, Qin, Cao, and colleagues (2019) studied volatile organic compounds in the surface water of the lake.

Case Study #5: Poyang Lake (鄱阳湖) Volatile Organics Study (2019)

The following carcinogenic and noncarcinogenic risks were calculated for the chemical contaminants in the lake. Dermal and ingested exposures were computed. The ten noncarcinogenic solvents and other contaminants arising from regional industry were identified and studied. Four were also known carcinogens.

Contaminants	Carcinogenic risk			Noncarcinogenic risk		
	Ingestion risk	Dermal risk	Total risk	Ingestion risk	Dermal risk	Total risk
1,1-dichloroethylene				1.28×10^{-7}	5.36×10^{-7}	1.28×10^{-4}
Methylene chloride	1.32×10^{-6}	5.55×10^{-9}	1.33×10^{-6}	1.10×10^{-1}	4.62×10^{-4}	1.11×10^{-1}
Cis-1,2-dichloroethylene				8.33×10^{-5}	3.50×10^{-7}	8.37×10^{-5}
Chloroform				1.75×10^{-2}	7.33×10^{-5}	1.76×10^{-2}
Benzene	7.60×10^{-7}	3.19×10^{-9}	7.63×10^{-7}	5.06×10^{-2}	2.13×10^{-4}	5.09×10^{-2}
Bromodichloromethane	5.76×10^{-7}	2.42×10^{-9}	5.79×10^{-7}	5.65×10^{-2}	2.37×10^{-4}	5.68×10^{-2}
Toluene				1.27×10^{-2}	5.33×10^{-5}	1.28×10^{-2}
Dibromochloromethane	8.24×10^{-8}	3.46×10^{-10}	8.27×10^{-8}	1.03×10^{-4}	4.32×10^{-7}	1.03×10^{-4}
Chlorobenzene				6.26×10^{-5}	2.63×10^{-7}	6.29×10^{-5}
Ethylbenzene				1.06×10^{-4}	4.45×10^{-7}	1.07×10^{-4}

Challenge: From the results of this analysis, identify any carcinogenic and noncarcinogenic contaminants where remedial action may be necessary, individually, combined by pathway, or all contaminants combined.

3.8.4 Case Study #6: Acrylamide

Acrylamide is produced in a natural reaction between carbohydrates and asparagine, an amino acid in plant-based foods, when exposed to high temperatures as in frying, roasting, and baking. High levels of acrylamide increase the risk of cancer in laboratory animals, and the topic has appeared in media frequently.

Case Study #6: Acrylamide: A Carcinogen in Baked/Frided Foods

Between 2011 and 2015, the US FDA (2022) sampled 2500 commonly available and popular starch-based food products to study acrylamide levels.

The following are a small selection of the food items (7) tested, together with the levels of acrylamide in parts per billion (ppb = μg/kg) as found in the FDA study.

Assumptions were lifetime exposures at the daily intake dose shown, by adult, 70 kg bw.

The EPA IRIS lists acrylamide: **RfD: 2×10^{-3} mg/kgbw·d**, and **SF: 5×10^{-1} mg/kgbw·d^{-1}**

Food item	ppb (μg/kg)	ppm (mg/kg)	Ingested per day	CDI mg/kg·d	Risk	Haz index
Granola bar	30	0.03	60 g	2.57E−05	1.28E−05	1.28E−02
Tortilla chips	40	0.04	30 g	1.71E−05	8.55E−06	8.55E−03
French fries	90	0.09	100 g	1.29E−04	6.45E−05	6.45E−02
Crackers	160	0.16	30 g	6.86E−05	3.43E−05	3.43E−02
Potato chip A	260	0.26	30 g	1.11E−04	5.55E−05	
Ginger cookies	1450	1.45	40 g	8.29E−04		
Potato chip B	2860	2.86	30 g			

Challenge:

1. Calculate the missing risk estimates and hazard index values. Were any of the exposures for carcinogens in excess of the "acceptable" level?
2. Were any of the noncarcinogens in excess of the "acceptable" level?
3. What interpretation and advice would you think appropriate for the general consuming public regarding these types of foods?

3.8.5 Case Study #7

Mercury in fish. Since the Minamata Bay incident, ingestion of organic mercury has become an urgent global concern. Questions such as this are common.

Case Study #7: Mercury in Fish

How much halibut can be consumed per week to avoid toxic amounts of mercury?

The question cannot be answered without a lot more information. Mercury (Hg) is a natural element found in all regions and media. Of the three forms (metallic, mercuric salts, and organic compounds), organic mercury, usually taken as methylmercury (CH_3Hg), is the most harmful.

Species	Mean ppm concentr	Max ppm concentr	Data source
Scallop	0.003	0.033	FDA 1991–09
Sardines	0.013	0.083	FDA 2002–10
Tilapia	0.013	0.084	FDA 1991–08
Salmon	0.014	0.086	FDA 1993–09
Herring	0.078	0.560	FDA 2005–12
Pickerel	0.096	0.310	FDA 1991–07
Tuna canned light	0.126	0.889	FDA 1991–10
Halibut	0.241	1.520	FDA 1992–09
Swordfish	0.995	3.220	FDA 1990–10

Mercury concentration varies enormously, up to three orders of magnitude between species and location. Mean concentrations are available from national and international sources. This is an extract of a chart from the US Food and Drug Administration (source: FDA(US)f 2022).

Hazard identification: Numerous surveys have shown that 50–75% of all mercury in fish is organic mercury, so we read "total" mercury as methylmercury.

Exposure assessment (CR): Convert weekly intake to daily intake.

Dose response: Methylmercury is listed (IRIS-EPA) with RfD 3.00E−04 (mg/kg·d).

Challenge: [A]: What is the hazard index for an adult (70 kg) who eats 350 g halibut each week? Assume halibut sold in the region contains mean mercury concentration of 1.00 ppm (1 mg/kg). **[B]:** Characterize the risk and advise the person appropriately.

3.8.6   Case Study #8

Application of newsprint to farmland. The recycling of organic materials to agricultural fields has been proposed as a way to reduce landfill needs as well as soil improvement. But the details need careful examination.

Case Study #8: Application of Newspaper to Farmland

A farmer intends to apply a mulch of shredded newspaper to his land, prompting a human health risk assessment. (Edited from JA Bukowski 1991)

Hazard identification: Ink: The only potentially hazardous constituent is carbon-black that contains several polycyclic aromatic hydrocarbons (PAHs) suspected of being carcinogenic. Paper/Newsprint: Considered free from hazards with one exception: dioxin (10 ppt) This is a B2 (possible) human carcinogen. The assessment will be for these two identified carcinogens.

Dose-response data: SF$_{(PAH)}$: $(11.5 \text{ mg/kg} \cdot \text{d})^{-1}$ **SF$_{(DIOXIN)}$:** $(1.56\text{E} + 05 \text{ mg/kg} \cdot \text{d})^{-1}$

Exposure assessment: Oral exposure through vegetables grown on the land, groundwater, meat, and products of animals grazing on the land. EPA estimates the average N. American eats 205 g vegetables/d (see note below). Proposed application: max 200 tons newspaper per acre (= 182,000 kg paper, containing 4900 kg ink, per acre).

Exposure from ink: At 200-ton paper/acre = Concentration of carbon-black in soil: 550 mg/kg, and 0.1% carbon-black is PAH, then **conc. in soil** is 0.55 mg/kg. **Bioconcentration factor (BCF)** is 0.06, so **Conc. in plants =** 0.55 mg/kg × 0.06 = 0.03 mg/kg. Adult vegetarian consuming 205 g veg/day **Intake:**	**Exposure from paper:** At 200 ton/acre Dioxin concentrations: 10 ng/kg (ppt).
$$\frac{\left(0.03\,mg\,/\,kg\right)\left(0.205\,kg\,/\,d\right)}{70\,kg}$$ $$= 9 \times 10^{-5}\,mg\,/\,kg \cdot d$$	$So: 181{,}818\,kg \times 10\,ng\,/\,kg$ $= 1.82 \times 10^{6}\,ng$ or $376\,ng\,/\,m^{2}$ **Dioxin conc. (soil)** will be 1.58 ng/kg (ppt) **BCF** is 0.01, so, **plant conc.** approx.: 1.58×10^{-5}ng/kg or 1.58×10^{-8}mg/kg · d **Intake:**
Excess cancer risk (ink): $(9 \times 10^{-5}$mg/kg · d$) \times (11.5$mg/ kg · d$)^{-1} = \mathbf{1 \times 10^{-3}}$	$$\frac{\left(1.58 \times 10^{-8}\right)\left(0.205\,kg\,/\,d\right)}{70\,kg}$$ $$= 4.6 \times 10^{-11}\,mg\,/\,kg \cdot d,$$ **Excess cancer risk (newspaper):** $(4.6 \times 10^{-11}$mg/ kg · d$) \times 1.56 \times 10^{5} = \mathbf{7.2 \times 10^{-6}}$

<u>Note:</u> The EPA's estimated intake of 205 g vegetables per day (cited in the report) appears remarkably low, even for N. American diet. We suspect the actual consumption is realistically much higher, but this can be adjusted at the end of the calculation for intake and risk.

Challenge: The intake and risk calculations have been completed based on the assumptions given.

[A] How would you characterize the carcinogenic risk from these two substances through consumption of 205 g vegetables grown on this soil every day?

[B] This proposal requires a decision. Based upon (A), what suggestions would you make at this plan approval stage to avoid exceeding acceptable risk levels?

3.8.7   Case Study #9

PCB spill on the road in Northern Ontario. Polychlorinated biphenyls (PCBs) were manufactured domestically from 1929 until manufacturing was banned in 1979. They have a range of toxicological effects and can harm animal, human, and environmental health.

Case Study #9: PCB Spill on Road in Northern Ontario

Several years ago, a tank truck carrying waste PCBs to a disposal site in Alberta passed through several Northern Ontario communities. A leak developed and went unnoticed by the driver for several miles until a couple following the truck in their car managed to indicate to him that something was dripping onto the road. As a result, the health unit and the Ministry of the Environment ordered a large-scale clean-up, including removal of some of the road surface. Concern was expressed for occupants of cars that had followed the vehicle, as well as pedestrians, cyclists, animals, and ultimately, road workers who removed the road surface.

Challenge: Identify, where appropriate, **source, release mechanism, transport mechanisms/media, transfer mechanisms, transformation mechanisms, exposure points, receptors,** and **exposure routes.**

3.8.8   Case Study #10: Benzene in Domestic Water Due to Industry

Benzene is one of the top 20 chemicals used by industry (in terms of volume) for the production of polymers, resins, lubricants, rubber, drugs, and detergents. It has numerous short-term effects and the US Department of Health and Human Services (DHHS) has established that benzene causes cancer and leukemia in humans.

Case Study #10: Benzene in Water Due to Proximity to Industry

A population of 100,000 living and working in a valley surrounded by oil refineries and petrochemical industries has been found to have, on average, **an individual daily intake of 140 μg (0.14 mg)** of the carcinogen <u>benzene</u>. Estimate the additional lifetime cancer cases expected for this population. The slope factor (SF) for benzene is 0.029 $(\text{mg/kg} \cdot \text{day})^{-1}$.

Note: The daily intake per person has been calculated but it is for the whole person. The first step is to convert intake to "per kg body weight." Assume the exposure has been constant for many decades and is expected to continue, so use lifetime of 70 years, in which case the EDxEF/AT cancel and the calculation is simplified to become:

$$\text{Intake}, (CDI) = \frac{(0.14\,\text{mg / day})}{(70\,\text{kg})} = 0.002\,\text{mg / kg} \cdot \text{d}$$

Incremental lifetime risk of cancer death = (CDI) × (SF) = $(0.002) \times (0.029) = 0.000058$

(Note: this clearly exceeds de minimis by 58 times).

Challenges: [A] Predict in this population the number of additional cancer deaths due to this benzene contamination on the hypothetical basis of it continuing unchanged.

[B] Assume that all benzene contamination has stopped, a remedial clean-up is taking place, and the community begins a new water supply this week. Use the 30-year average exposure (ED) on the basis that some people have been exposed their whole life, while others arrived only in the last year. The AT remains at 25,550 days because benzene is a carcinogen. Predict the total additional cancer deaths due to this exposure.

3.8.9   Case Study#11

Residential water contamination by metal plating works. Metal smelting, refining, and plating often cause harmful environmental contamination, especially in an urban setting.

Case Study #11: Water Contam. by Metal Plating Works

Hazardous waste from a metal plating facility has contaminated groundwater. A nearby community's water supply has been shown to contain cyanide at a conc. of 350 µg/L, nickel at 1200 µg/L, and chromium (III) at 12,400 µg/L. Do these exposures indicate an unacceptable hazard either separately or combined?

Assuming the average daily water intake is 2 L and the body weight of an average human is 70 kg, the noncarcinogenic lifetime average daily dose is (using cyanide as the example):

$$\mathbf{LADD}\,(\text{cyanide}) = \frac{0.350\,\text{mg}\,/\,\text{L} \times 2\,\text{L}\,/\,\text{day}}{70\,\text{kg}}$$

$$= 1.00 \times 10^{-2}\,\text{mg}\,/\,\text{kg} \cdot \text{day}$$

The hazard index: $(I_N\,/\,\text{RfD}) = 0.010\,/\,0.020 = 0.5000$

Substance	Conc (mg/L)	I_N (LADD) (mg/kg·day)	RfD	Hazard index (dose/RfD)
Cyanide	0.350	1.00E–02	2.00E–02	0.500
Nickel soluble salts	1.200	3.43E–02	2.00E–02	_____
Chrome (III)	12.400	_____	1.00E+00	_____
				_____ = (tot)

Note: This is a simple additive model. Strictly speaking, additivity assumes that each agent has a similar detrimental effect on the body. It is justified, for example, when dealing with mixed exposure to cadmium, mercury, uranium, and chromium, all of which affect the kidneys. However, adding the hazard indices for separate compounds in the manner demonstrated here may overestimate the risk if they induce toxic effects by different mechanisms and affect different organ systems. In this example, the compounds present in the mixture do not affect a common target organ, or act by a common mechanism, and an argument can be made for segregating the findings by critical effect, with separate hazard indices being reported on an organ-specific basis. This should only be done when the mechanisms of the non-carcinogenic toxicants are well known, and the target organs understood. When in doubt, the opinion of a toxicologist should be sought.

Challenge: Complete the results for the other two materials, and summarize the RfD for combined contaminants as well as separately.

3.8.10 Solutions for Case Studies #3 to #11

Solutions for Case Studies #3 to #11

Case #	
3	[A] The arsenic content in this water is nearly 50 times above acceptable level [B] Drinking 2000 ml/50 = 40 ml per day would reach HI of 1.0. [C] The EPA's *acceptable level* of arsenic in water (10 µg/L or 0.01 mg/L) results in a daily intake of 0.000286 mg/kg·d $\text{Haz.Index} = \dfrac{I_N}{RfD} = \dfrac{0.000286}{0.0003} = 0.952$ HI is very close to 1.0, as expected. Decision to recall the product by EPA was appropriate.
4	None of the options (including the "No action" option) predict excess cancer deaths above the current de minimis 1×10^{-6}.
5	The dermal pathway contributes negligble risk (carcinogens or noncarcinogens). The **noncarcinogenic** contaminants do not represent a concern. The combined HI for methylene chloride (both pathways) is highest at approximately 1.1. The total HI for all other noncarcinogens combined remains less than 1.0. For **carcinogens**, methylene chloride is the only single contaminant reaching de minimis (1.33×10^{-6}) by ingestion. The other three carcinogenic contaminants combined (by ingestion) reaches HI of 1.4×10^{-6}. The carcinogenic risk is of marginal concern and recommendation would be to repeat the survey to explore the trend.
6	<table><tr><td></td><td>CDI</td><td>Risk</td><td>HI</td></tr><tr><td>PCA</td><td></td><td></td><td>0.0555</td></tr><tr><td>G-C</td><td></td><td>4.145E-04</td><td>0.4145</td></tr><tr><td>PCB</td><td>1.226E-03</td><td>6.129E-04</td><td>0.6129</td></tr></table> 1. ALL products exceeded de minimis cancer risk of 1×10^{-6}. 2. None exceeded hazard index of 1.0 3. Be aware, reduce, prefer lighter color-baked products.

7	Average consumption (CR) of 350 g/w (= 50 g/day) and maximum concentration of ingested mercury in fish as 1.0 mg/kg:

$$I_N = \frac{[1.0\,\text{mg}/\text{kgF}] \times (0.05\,\text{kg}/\text{d})}{70\,\text{kg}} = 7.1 \times 10^{-4}\,\text{mg}/\text{kg} \cdot \text{d}$$

$$HI: \quad \frac{I_N}{RfD} = \frac{7.1 \times 10^{-4}\,\text{mg}/\text{kg} \cdot \text{d}}{3.0 \times 10^{-4}\,\text{mg}/\text{kg} \cdot \text{d}} = 2.4$$

Characterization: Given these assumptions, reduction of intake is recommended. Options include limiting halibut consumption to 145 g/week (350 g/2.4) or selecting a species with lower mercury concentration. Smaller species contain much less mercury, but local and national health and food safety agencies should be consulted about fish in your area. Pregnant women, infants, children, and lactating mothers are advised to be extra careful, such as eating only low-mercury species. Interpretation should also balance the considerable benefits of marine foods, especially oils.

8	[A] Carcinogenic risk from paper is close to being acceptable (7.2×10^{-6}), but risk from PAH in ink (1×10^{-3}) exceeds de minimis by 1000 times.

[B] As the two are combined in the same material, we consider only ink risk. For acceptable risk, divide 200 tons/1000 = 0.2 ton (200 kg) per acre application.

This is probably not cost-effective or useful. Also, if the 205 g veg/day is increased, it will further increase risk.

Recommend: not approving the proposal.

9	**Source:** Tank on a truck. **Release mechanism:** Leak in tank.

Transport mechanisms: Moving truck. **Media:** Droplets/vapor in moving air. **Transfer mechanisms:** Onto road surface, onto vehicle surfaces.

Transformation mechanisms: None (chemically stable).

Exposure points: In car following truck; working on road surface, pedestrians, children on bicycles, disposal area exposure? (unknown)

Receptors: Occupants of car (following).

Exposure routes. Inhalation, transdermal, ingestion?

(continued)

Case #	
10	[A] In a population of 100,000 = (5.8E−05) × (1.00E+05) = 5.8 (approx. 6) additional cases [B] If we use 30 y ED instead of 70 y, the intake will be 0.000857 and the risk will be 0.0000249, or 25 additional lifetime deaths in this population
11	The hazard index combined is 2.6. The Ni content by itself exceeds 1.0. If the Ni could be removed, the Cya and Cr might be acceptable (HI<1.0). Check with toxicologist which organ systems are targets for these substances, and whether other environmental sources should be considered.

Table for Case 11:

	Conc mg/L	Intake mg/kg·d	RfD mg/kg·d	HI
Cya	0.350	0.010	0.02	0.500
Ni	0.12	0.034	0.02	1.715
Cr	12.4	0.354	1.00	0.354
			Tot:	**2.569**

Practice Exercises: Chapter 3

	Assumptions and info

1. Propylene oxide
 	
Propylene oxide (C3H6O) is used in many industrial applications and formulations. You are conducting a risk assessment for workers who have been exposed to airborne propylene oxide vapor at a mean concentration of **10 mg/M³** air. The plant has been operating for 5 years, and the workers are present during an 8-hour shift each day for usually 220 days per year. Estimate:
 	
(a) The incremental lifetime carcinogenic risk
 	
(b) The noncarcinogenic hazard index
 	
(c) What highest single concentration in air would just meet BOTH the "acceptable" risks? (carcinogenic and noncarcinogenic)
	BW: 70 kg
	Resp rate (CR): 0.95 M³ per hour
	EPA IRIS SF for propylene oxide (inhal):
	  2.60E–01 (mg/kg·d)⁻¹
	EPA IRIS RfD (inhal) for propylene oxide:
	  8.58E–03 (mg/kg·d)

2. Carbon tetrachloride in workplace During the production of **carbon tetrachloride** (CTC), workers are exposed to CTC vapors during an interval of 3 hours a day at a concentration in air of **0.1 mg/M³**. Assess the carcinogenic risk for a worker who is exposed in this way for 5 years.	Assumptions and info **Assumed respiration rate:** 1.2 M³/hr **BW** workers: 70 kg **EPA IRIS SF** for carbon tetrachloride (inhalation): 1.30E–01 (mg/kg·d)⁻¹
3. Chem lab vapors A university chemistry lab has several old containers of toluene in a storage room that were not closed tightly. Several graduate student researchers have been working in this lab for an average of 5 hours a day for 6 days a week. This has been continuing for 6 months. Air sampling has shown that the concentration of toluene in the laboratory air is around **0.5 mg/M³**. Calculate the noncarcinogenic risk ("hazard index") using the information.	Assumptions and info **BW:** 70 kg **CR** (resp rate): 0.85 M³/hr **IRIS RfD** toluene (inhal): 1.40E+00 (mg/kg·d)

(continued)

4. Dry-cleaning operation

	Assumptions and info
A dry-cleaning operation operates for 9 am to 6 pm 6 days a week. An inspector finds that a return tube from the condenser stage has broken and that the air in the shop carries a strong odor of the dry-cleaning solvent **tetrachloroethylene** (also known as **perchloroethylene**, $Cl_2C=CCl_2$). The owner-operator doesn't notice the odor (a common problem with longer-term exposure). The air is sampled and found to have a concentration of **2.0 mg/M³**.	CR (resp rate): 0.9 M^3/hr BW: 70 kg EPA IRIS RfD: 1.00E–02 $(mg/kg \cdot d)$
The exposure is thought to have begun 9 months ago when the condenser was damaged but not repaired. For the operator's 9-month exposure: **(i)** Estimate the chronic daily intake. **(ii)** Estimate the noncarcinogenic risk ("hazard index"). **(iii)** What should be the maximum concentration during working hours so as not to exceed an HI of 1.0?	

5. PCB in water

	Assumptions and info
A village water supply has been fed by wells drilled into an aquifer but is now been found to have been contaminated by leaking transformer fluid (**arachlor** a **polychlorinated biphenyl, PCB**) from equipment buried in 1950 on a former industrial site adjacent to the village. The mean concentration in drinking water of **arachlor** is close to **0.1 µg/liter.**	CR: 2 L/d BW: 70 kg EPA SF PCBs: 3.5E–01 $(mg/kg \cdot d)^{-1}$ (ingestion)
Assuming this water is consumed from this source for a lifetime, what is the predicted incremental carcinogenic risk for an adult under these conditions?	

	Assumptions and info
6. Malathion on vegetables	CR: 500 g vegetables/day
Residents in a rural area of Ontario has been found to have been exposed to a pesticide through vegetables sold in a popular farmer's market. A local farmer has been using malathion in a manner that leaves a residue on a wide range of leafy vegetables. The mean concentration is **3 mg/kg of greens** (3 ppm by weight). The EPA has no carcinogenic risk data but lists malathion with a reference dose (RfD) of 2.0E−02 (2.0×10^{-2}) mg/kg · d. What is your judgment for adult exposure on an ongoing basis?	ED every day EF 70 yrs BW: 70 kg
7. Methanol in whisky	Assumptions and info
Acute **methanol** exposure (high concentrations over a short period) can damage the optic nerve within hours. Here we are concerned with chronic methanol exposure. Thus, in the case of methanol risk assessment, we are finding the hazard index from exogenous exposure (exposure from a source outside the body) that adds to background levels of methanol derived from normal diet. The EPA RfD is **0.5 mg/kg · day.** A person drinks 1 fluid ounce of homemade whisky each day. It contains **0.1 percent methanol.** Estimate the hazard index for this continuing exposure.	Assume 1 fl oz has mass of 26 g and also assume that 1000 mg = 1 g = 1 ml. BW: 70 kg $EF \times ED = AT$ (it cancels) For this type of "all-day, every day for life" scenario, the calculation becomes $$\frac{[C] \times CR}{BW}$$
8. Worker exposed to formaldehyde	Assumptions and info
A female (54 kg) worker in a pressed-wood manufacturing plant breathes air containing formaldehyde vapor (from the hot adhesive) with a mean concentration of **15 µg/cubic meter of air**. The shift is 7 hrs a day, approx. 230 days/year over a 5-year period. What is the carcinogenic risk for this exposure?	Respiration rate 0.90 M3/hr EPA/IRIS CPF/SF for formaldehyde (inhaled): **4.5E−02** $(mg/kg \cdot d)^{-1}$

(continued)

	Assumptions and info
9. Rural school with solvent in water The water supply for a rural school has been found to be contaminated by a solvent (**dibromochloromethane**) from a nearby electronics factory. The mean concentration of the solvent is 100 μg/L (**0.1 mg/L**). This contamination is assumed to have begun following the opening of the factory 18 months ago. Carry out (a) an assessment for carcinogenic risk acquired during this period, (b) the hazard index, and (c) advise the school board of your findings.	BW: 29 kg EF:190 days/yr EPA/IRIS RfD: (oral) 2.00E−02 (mg/kg · d) EPA/IRIS CPF: (oral) 8.40E−02 (mg/kg · d)$^{-1}$ CR (during school day): 2 L/d
10. Parathion on lettuce	Assumptions and info
This "contact" weed killer is showing up on local lettuce at the rate of **400 μg per kg** lettuce. Assume a person eats **200 g** lettuce/day for life. Calculate the haz index.	BW: 66 kg EPA RfD (ingest) for parathion is 4.5E−03 (mg/kg · d).
11. Dichlorvos in water	Assumptions and info

What is your estimated lifetime cancer risk for an adult who is exposed at home to drinking water containing 0.001 mg/L dichlorvos. The exposure is for 2 years. She is employed as a grade 3 teacher at a school not affected by the contaminated water.
EPA SF dichlorvos: 2.9E−01 (mg/kg · d)$^{-1}$
	CR = 2 L/d
	BW = 55 kg
	AT = (carc.) 25.550d
	EF = This is time NOT at school during the year. Calculate # days NOT at school using weekly units (39 wks. school year, 13 wks. vacation)
	ED = 2y

	Assumptions and info
12. Mine worker/nickel dust	CR: 1.2 M^3/h
You are assisting in the assessment of a worker in a nickel mine. The mean concentration of **nickel refinery dust** in air is 2.0 microgram/M^3. The workers are at work about 200 days a year in single 8 hr shifts. They have been at this work for 8 years.	BW = 70 kg
	IRIS SF Nickel (inhaled): 8.4×10^{-1} (mg/kg · d)$^{-1}$
Has the typical worker been exposed to excessive nickel? Calculate the additional risk of death from cancer. What recommendations, if any, would you make to the company to reduce the risk for future workers?	
13. MEK at workplace	Assumptions and info
Workers exposed to methyl ethyl ketone at an average concentration in air of 16 mg/m^3 during daily 8 hr shifts, 220 days a year for two years. Assume respiration at 1.2 M^3/hr. Assume 80% retained dose.	• CPF: nil
	• Inhal RfD: 0.286 mg/kg · d
	• Retained: 0.8
	• BW: 70 kg
14. Chlorine dioxide in flour mill	Assumptions and info
Assess risk or HI for temporary flour mill workers employed in bleaching process using ClO_2. Assume exposure takes place 4 hr each day, 5d/w, at concentration of approximately 4 µg/M^3. They are employed for 6 months.	• CPF: no data
	• Inhal. RfD: 5.72E−05 mg/kg · d
	• CR: 1.2 M^3/h
	• 5d/w = 21.7d/m (use either)

(continued)

	Assumptions and info.
15. Thallium sulfate in well water	
The water supply to a rural elementary school has been found to have thallium sulfate concentration at 2 µg/L. This is both PH issue and OH&S issue. Assess the hazard index for _one school year_ for: (a) Child 16 kg bw (b) Child 29 kg bw (c) Adult staff 70 kg bw (Note this is a PH and OH&S issue)	• Oral RfD. 8.0E−05 mg/kg · d • CR: 1 L/d for 16 kg kids • CR: 1.5 L/d for 29 kg kids • CR: 2.0 L/d for adults • EF × ED 200 d for kids • EF × ED 210 days for adults • AT 9 m (274 d) • Assume all daily intake is drunk at school during school days
16. Arsenic in population's water	
Arsenic (inorganic) occurs naturally in groundwater in some parts of the world. It causes a wide range of chronic (long-term) health effects. Make an analysis of the risks that can be expected over the lifetime of people in such an area where the concentration of arsenic in the water supply is 10 µg/L (0.01 mg/L). (a) What is the (lifetime) risk of cancer due to arsenic in drinking water for a member of this population? (b) What is the (lifetime) hazard index due to arsenic in drinking water for a member of this population? (c) Does the cancer risk exceed the de minimis risk level? If yes, by how many times? (d) Does the HI exceed the "acceptable" level? If yes, by how many times? (e) How many lifetime incremental arsenic-related cancer deaths would you expect in a region with a population of 540,000? (f) To render this water "acceptable" from _both_ cancer and noncancer risk, given that the daily intake must remain at 2 L/day, what should the concentration of arsenic in this water _not exceed?_	Mean water intake: 2 L/d Mean body weight: 62 Kg ED: 70 y EF: 365 d/y $AT_N = 70y \times 365d$ $AT_C = 70y \times 365d$ ABS: assume 100% (use 1.0) RR: assume 100% (use 1.0) SF: 1.75 $(mg/kg \cdot d)^{-1}$ RfD: 0.0003 $(mg/kg \cdot d)$

	Dioxane RfD: no data
	Dioxane SF:
	$1.10E{-}02\ (mg/kg \cdot d)^{-1}$

17. 1,4-dioxane levels in drinking water

A small community is served by a drinking water system connected to a well.
Industrial waste buried in the ground is suspected to be the source of the
1,4-dioxane which is showing up in the water system at around 100 µg/L.
**Assess the incremental (lifetime) carcinogenic fatality risk in the following
scenarios, together with their remedies.**

(a) Adult, continuous exposure at 70 kg bw. 2 L/d

(b) How many times is this risk greater than the "acceptable" (de minimis)
level of 1.0E−06?

(c) One remedy has been proposed that would supply half their assumed water
daily intake (or 1 L/d per person) as bottled water, with the rest from the
community supply.

(d) An alternative proposition is to add ion-exchange resin filters to the water
supply that would drop the concentration of dioxane to 40 µg/L.

(e) Which solution (c) or (d) would present the lower risk?

(f) Regardless of which option, what is the maximum concentration of
1,4-dioxane in water so as not to exceed the 1.0E−06 risk level for this
lifetime scenario?

(continued)

Ques	Detailed solutions and comments
1	Two calculations here: I_N and I_C.
	$I_N = \dfrac{\left[10\,mg\,/\,M^3\right]\left(0.95M^3\,/\,h\right)\left(8h\,/\,d\right)\left(220d\,/\,y\right)\left(5y\right)}{\left(70\,kg\right)\left(5 \times 365d\right)} = 0.6544\,mg\,/\,kg \cdot d$ HI : $\dfrac{0.6544}{0.00858} = 76.27$
	$I_C = \dfrac{\left[10\,mg\,/\,M^3\right]\left(0.95M^3\,/\,h\right)\left(8h\,/\,d\right)\left(220d\,/\,y\right)\left(5y\right)}{\left(70\,kg\right)\left(25{,}550d\right)} = 0.04674\,mg\,/\,kg \cdot d$
	(risk) : $0.04674 \times 0.26 = 1.22E-02$
	HI exceeds acceptable level by 76 times, while incremental risk is over twelve thousand times greater than 1.0E−06. High priority. "Stop-work order" is appropriate. Limiting concentration in breathable air to **8.2×10^{-4} mg/M³ (or 0.82×10^{-1} mg/M³)**. This will reduce <u>both</u> risk and HI to acceptable levels
2	
	$I_C = \dfrac{\left[0.1\,\mu g\,/\,M^3\right]\left(1.2M^3\,/\,h\right)\left(3h\,/\,d\right)\left(225d\,/\,y\right)\left(5y\right)}{\left(70\,kg\right)\left(25{,}550d\right)} = 0.000226\,mg\,/\,kg \cdot d$
	(risk) : $0.000226 \times 0.13 = 2.94E-05$
	The incremental carc. risk is about 30 times more than acceptable
3	
	$I_N = \dfrac{\left[0.5\,mg\,/\,M^3\right]\left(0.85M^3\,/\,h\right)\left(5h\,/\,d\right)\left(156d\,/\,y\right)\left(1y\right)}{\left(70\,kg\right) \times \left(182d\right)} = 0.02620$ **HI :** $\dfrac{0.02602}{1.4} = \mathbf{0.019}$
	The hazard Index is well below 1.0, so no further action is needed. (EFxED): 6 months (26 weeks) at 6 days a week = 156 d

4

$$I_N = \frac{\left[2\,mg\,/\,M^3\right]\left(0.9\,M^3\,/\,h\right)\left(9\,h\,/\,d\right)\left(6\,d\,/\,w\right)\left(52\,w\,/\,y\right)\left(0.75\,y\right)}{\left(70\,kg\right)\times\left(0.75\right)\left(365\,d\right)} = 0.19782 \quad \mathbf{HI}: \frac{\mathbf{0.19782}}{\mathbf{0.01}} = \mathbf{19.78}$$

(i): IN = 19.78 **0.01**
(II) HI 19.78, nearly 20 times higher than acceptable
(III) Accepted concentration of perchloroethylene $(2\,mg/M^3)/19.8 = 0.10\,mg/M^3$

5

$[0.1\,\mu g/L = 0.0001\,mg/L]$ and ED.EF cancels AT:

$$I_C = \frac{\left[0.0001\,mg\,/\,L\right]\left(2\,L\,/\,d\right)}{\left(70\,kg\right)} = 0.00000285\,mg\,/\,kg\cdot d$$

$$(\text{risk}): \mathbf{2.85\times10^{-6}\times0.35 = 0.000000997}\left(\text{or}\,\mathbf{1\times10^{-6}}\right)$$

No further action is necessary.

6

$$I_N = \frac{\left[3\,mg\,/\,kg\right]\times\left(0.5\,kg\,/\,d\right)}{\left(70\,kg\right)} = 0.02143\,mg\,/\,kg\cdot d \quad \mathbf{HI}: \frac{\mathbf{0.02143}}{\mathbf{0.02}} = \mathbf{1.07}$$

Close to acceptable. No further action.

(continued)

Ques	Detailed solutions and comments
7	This one needs some careful conversions to get the concentration to mg/L. First, we need to assume that 1 g = 1 ml. (It is not exactly true with alcohol).
	Note: 0.1 percent can be expressed as = 0.1 mg in 100 mg, or 1.0 mg in 1000 mg = 1 g, so 1000 mg in 1000 g (and if we assume 1 g = 1 ml), then 1000 mg in 1000 ml or 1 L
	$$I_N = \frac{[1{,}000\,mg\,/\,L]\,(0.026\,L\,/\,h)}{(70\,kg)} = 0.37143\,mg\,/\,kg \cdot d \quad \textbf{HI}: \frac{0.37143}{0.5} = 0.74 \;(\textbf{No further action})$$
8	$$I_C = \frac{\big[0.015\,mg\,/\,M^3\big]\big(0.9\,M^3\,/\,d\big)\big(7\,h\,/\,d\big)\big(230\,d\,/\,y\big)\big(5\,y\big)}{(54\,kgbw)\times(25{,}550\,d)} = 0.00007877\,mg\,/\,kg \cdot d$$
	$$(\textbf{risk}): 0.0000788 \times 0.045 = \textbf{3.54E}-\textbf{06}$$
	This is about 3.5 times higher than the de minimis level, and although this question did not require it, it can be calculated that a concentration of 4.2 μg/M^3 would satisfy the de minimis level.
9	$$I_N = \frac{\big[0.1\,mg\,/\,L\big]\big(2\,L\,/\,d\big)\big(190\,d\,/\,y\big)\big(1.5\,y\big)}{(29\,kg)\times(1.5)(365\,d)} = 0.003590\,mg\,/\,kg \cdot d \quad \textbf{HI}: \frac{0.003590}{0.02} = \textbf{0.1795}$$
	$$I_C = \frac{\big[0.1\,mg\,/\,L\big]\big(2\,L\,/\,d\big)\big(190\,d\,/\,y\big)\big(1.5\,y\big)}{(29\,kg)\times(25{,}550)} = 0.00007693\,mg\,/\,kg \cdot d$$
	$$(\textbf{risk}): 0.00007693 \times 0.084 = \textbf{6.5E}-\textbf{06}$$
	Carcinogenic risk is about 6.5 times higher than acceptable. Noncarc. hazard index is within acceptable range.

10

$$I_N = \frac{[0.4\,\text{mg}\,/\,\text{kg}](0.2\,\text{kg}\,/\,\text{d})}{(66\,\text{kgbw})} = 0.001212 \quad \textbf{HI}: \frac{0.001212}{0.0045} = \textbf{0.269 mg}\,/\,\textbf{kg}\cdot\textbf{d}$$

No action necessary.

11 Here, the exposure is during the time at home. EF is calculated in week segments:

Weekdays at home during 39 wk school year: 39 w (16 h/24 h)(5 d/w)	= 130 d	
Weekends at home during 39 wk school year: 39 w (2 d/w)	= 78 d	
3 months in year at home:	13 w (7 d/w)	= 91 d
Total day-equivalents at home for 1 year) = (the EF)		**= 299 d**

(As a cross-check, the actual school time is (8 h/24 h)(21 d/m)(9 m) = (65 d)
which completes the year: (364 d)

$$I_C = \frac{[0.001\,\text{mg}\,/\,\text{L}](2\,\text{L}\,/\,\text{d})(300\,\text{d}\,/\,\text{y})(2\,\text{y})}{(55\,\text{kg})\times(25{,}550)} = 8.51\times10^{-7}\ \text{mg}\,/\,\text{kg}\cdot\text{d}$$

$$(\textbf{risk}): \textbf{8.51}\times\textbf{10}^{-7}\times\textbf{0.29} = \textbf{2.46}\times\textbf{10}^{-7}$$

(Below de minimis level of incremental risk.risk)

(continued)

Ques	Detailed solutions and comments
12	$$I_C = \frac{\left[0.002\,\text{mg}\,/\,\text{M}^3\right]\left(1.2\,\text{M}^3\,/\,\text{h}\right)\left(8\,\text{h}\,/\,\text{d}\right)\left(200\,\text{d}\,/\,\text{y}\right)\left(8\,\text{y}\right)}{\left(70\,\text{kg}\right)\times\left(25{,}550\right)} = 0.00001718\,\text{mg}\,/\,\text{kg}\cdot\text{d}$$ $$(\textbf{risk}): \mathbf{0.00001718 \times 0.84 = 1.443E-05}$$ The lifetime cancer risk for these nickel refinery workers is about 14 times too high. The concentration should not exceed 0.143 μg/M³.
13	$$I_N = \frac{\left[16\,\text{mg}\,/\,\text{M}^3\right]\left(1.2\,\text{M}^3\,/\,\text{h}\right)\left(8\,\text{h}\,/\,\text{d}\right)\left(220\,\text{d}\,/\,\text{y}\right)\left(2\,\text{y}\right)\left(0.8\right)}{\left(70\,\text{kg}\right)\times\left(2\times365\,\text{d}\right)} = 1.0581\,\text{mg}\,/\,\text{kg}\cdot\text{d}$$ $$(\textbf{HI}): \mathbf{1.0581\,/\,0.286 = 3.70}$$ The MEK workplace concentration is between 3 and 4 times higher than the HI of 1.0.
14	$$I_N = \frac{\left[0.004\,\mu g\,/\,\text{M}^3\right]\left(1.2\,\text{M}^3\,/\,\text{h}\right)\left(4\,\text{h}\,/\,\text{d}\right)\left(5\,\text{d}\,/\,\text{w}\right)\left(52\,\text{w}\,/\,\text{y}\right)\left(0.5\,\text{y}\right)}{\left(70\,\text{kg}\right)\times\left(182\,\text{d}\right)} = 0.0001959\,\text{mg}\,/\,\text{kg}\cdot\text{d}$$ $$(\textbf{HI}): \mathbf{0.0001959\,/\,5.72E-05 = 3.425}$$ Work environment exceeds HI of 1.0 by 3.4 times.

15

16 kg kids : $I_N = \dfrac{\left[0.002\,\text{mg}/\text{L}\right]\left(1\,\text{L}/\text{d}\right)\left(200\,\text{d}/\text{y}\right)\left(1\,\text{y}\right)}{\left(16\,\text{kgbw}\right)\times\left(274\,\text{d}\right)} = 0.00009124$

$$(\mathbf{HI}) : \mathbf{0.00009124}\,/\,\mathbf{8.0E-05} = \mathbf{1.14}$$

29 kg kids : $I_N = \dfrac{\left[0.002\,\text{mg}/\text{L}\right]\left(1.5\,\text{L}/\text{d}\right)\left(200\,\text{d}/\text{y}\right)\left(1\,\text{y}\right)}{\left(29\,\text{kgbw}\right)\times\left(274\,\text{d}\right)} = 0.0000755$

$$(\mathbf{HI}) : \mathbf{0.000766}\,/\,\mathbf{8.0E-05} = \mathbf{0.94}$$

Adults : $I_N = \dfrac{\left[0.002\,\text{mg}/\text{L}\right]\left(2.0\,\text{L}/\text{d}\right)\left(210\,\text{d}/\text{y}\right)\left(1\,\text{y}\right)}{\left(70\,\text{kgbw}\right)\times\left(274\,\text{d}\right)} = 0.0000438$

$$(\mathbf{HI}) : \mathbf{0.0000247}\,/\,\mathbf{8.0E-05} = \mathbf{0.55}$$

All below 1.0. Note: the larger body weight reduces the HI. Kids may be at greater risk and should be given access to alternate water supply.

(continued)

Ques	Detailed solutions and comments
16	$$I_C = \frac{[0.010\,\text{mg}/\text{L}](2\,\text{L}/\text{d})}{(62\,\text{kgbw})} = 0.0003226\,\text{mg}/\text{kg}\cdot\text{d}$$ $$(\textbf{risk}): \textbf{0.0003226} \times \textbf{1.75} = \textbf{0.0005645 or 5.64E} - \textbf{04}$$ (a) $$I_N = \frac{[0.010\,\text{mg}/\text{L}](2\,\text{L}/\text{d})}{(62\,\text{kgbw})} = 0.0003226\,\text{mg}/\text{kg}\cdot\text{d}$$ $$(\textbf{HI}): \textbf{0.0003226} / \textbf{0.0003} = \textbf{1.075}$$ (b) (c) Cancer risk exceeded by 564 times. (d) HI acceptable (e) 305 extra lifetime deaths from this population (f) 0.017 µg/L
17	a: 3.14E–05 b: 31.4 times c: 1.57^{-5} d: 1.257E–05 e: (d) f: 3.2 µg/L

References

Ames B. Nature's chemicals and synthetic chemicals – comparative toxicology 3. Proc Natl Acad Sci. 1990;87:7782.

Bollaerts K, Aerts M, Faes C, Grijspeerdt K, Dewulf J, Mintiens K. Human salmonellosis: estimation of dose-illness from outbreak data. Risk Anal. 2008;28(2):427–40.

Bukowski JA. Cancer risk from the application of newsprint to farmland. In: Garrick BJ, Geckler WC, editors. The analysis, communication, and perception of risk. New York: Plenum Press; 1991.

Canada Government. Federal contaminated site risk assessment in Canada, Part II: Health Canada Toxicological Reference Values (TRVs) and chemical-specific factors, Version 2.0. Queens Printer, Ottawa, 2013. Accessed Nov 2022.

Cheasley R, Keller CP, Setton E. Lifetime excess cancer risk due to carcinogens in food and beverages: urban versus rural differences in Canada. Can J Public Health. 2017;108(3):e288–95.

EPA(US)a. Use of Monte Carlo simulation in risk assessments: Region 3 technical guidance manual: risk assessment. Hazardous Waste Management Division, Office of Superfund Programs. Philadelphia. EPA 903-F-94-001 - February 1994. Accessed 17 Sept 2022 at: https://www.epa.gov/risk/use-monte-carlo-simulation-risk-assessments.

EPA(US)b. Integrated Risk Information System. (Database) US Environmental Protection Agency. Last updated 2022. Accessed 2022 07 06 at: https://www.epa.gov/iris.

EPA(US)c. Superfund reports: Lackawanna refuse. Old Forge, Pa. 2019. Accessed July 2022 at: https://cumulis.epa.gov/supercpad/cursites/csitinfo.cfm?id=0301220.

EPA(US)d. Guidelines for carcinogen risk assessment /630/P-03/001F. Washington, DC: U.S. Environmental Protection Agency; March 2005. Accessed 16 Sept 2022 at: https://www.epa.gov/sites/default/files/2013-09/documents/cancer_guidelines_final_3-25-05.pdf.

EPA(US)e. Exposure factors handbook 2011 edition (final). Washington, DC. EPA/600/R-09/052F: U.S. Environmental Protection Agency; 2011.

FDA(US)f. Mercury Levels in Commercial Fish and Shellfish (1990–2012) Updated Feb/2022. Accessed Nov 2022 at: https://www.fda.gov/food/metals-and-your-food/mercury-levels-commercial-fish-and-shellfish-1990-2012.

EPA (US)g. Guidelines for carcinogenic risk assessment. EPA/630/P-03/001F. Washington, DC: United States Environmental Protection Agency. Risk Assessment Forum; March 2005.

EPA (US)h. Guidelines for carcinogenic risk assessment. Washington, D.C. September 24, 1986. Federal Register (51 FR 33992-34003).

Evans JS, et al. A distributional approach to characterizing low dose cancer risk, 14. Risk Anal. 1992;25:31.

Health Canada. Final human health state of the science report on lead. Ottawa: Minister of Health; 2013. Accessed July 2022

Keenan RE. Risk Anal. 1994;14(3):225–30.

National Research Council. Risk assessment in the Federal Government: managing the process. Washington, DC: The National Academies Press; 1983. https://doi.org/10.17226/366.

Qin P, Cao FM, Lu SY, et al. Occurrence and health risk assessment of volatile organic compounds in the surface water of Poyang Lake in March 2017. Royal Society of Chemistry Adv. 2019;9:22609–17.

ProMED-mail. http://www.promedmail.org. International Society for Infectious Diseases. Wed 7 Mar 2007. Source: FDA (US Food and Drug Administration) News, press release, 2007.

Swirsky-Gold L, Backman GM, Hooper NK, Peto R. Ranking the potential carcinogenic hazards to workers from exposures to chemicals that are tumorigenic in rodents. Environ Health Perspect. 1987;76:211–9.

Thompson KM, et al. Monte Carlo techniques for quantitative uncertainty analysis in public health risk assessments, 12. Risk Anal. 1992;153:55–6.

Travis CC, Hester ST. Background exposure to chemicals: what is the risk? Risk Anal. 1990;10(4):463.

US Environmental Protection Agency. Ecological effects test guidelines OCSPP 850.1730: Fish bioconcentration factor (BCF). EPA 712-C-16-003. Office of Chemical Safety and Pollution Prevention. Washington, DC; 2016.

Vallero DA. Environmental biotechnology: a biosystems approach. In: Environmental risks of biotechnologies. Cambridge, MA: Academic Press; 2010. p. 325–400.

Wilson R, Jones-Otazo H, Petrovic S, Mitchell I, Bonvalot Y, Williams D, Richardson GM. Revisiting dust and soil ingestion rates based on hand-to-mouth transfer. Hum Ecol Risk Assess. 2013;19(1):158–88.

World Health Organization. Lead poisoning. 2022. https://www.who.int/news-room/fact-sheets/detail/lead-poisoning-and-health. Accessed Feb 2023.

Xie G. A novel Monte Carlo simulation procedure for modelling COVID-19 spread over time. Sci Rep. 2020;10:article #13120.

Qualitative Risk Assessment Methods

4

Abstract

Although the primary "risk assessment" focus of this book addresses quantitative methods, several qualitative methods of risk analysis are widely used for evaluation and decision-making across all industrial, institutional, educational, and commercial organizations.

Qualitative analysis methods describe risks in relative terms, without introducing complex calculations or requiring computing of probabilities. Numbers are sometimes used but only to rank a risk as being relatively higher or lower than another risk, or establishing a priority for remediation or resolution.

This chapter offers an introduction with examples and applications for the following qualitative processes.

- Preliminary Risk Analysis (PRA)
- Failure Mode Effects Analysis (FMEA)
- Fault tree analysis (FTA)
- Management Oversight and Fault Tree (MORT)
- Hazard and Operability (HAZOP)

Hazard Analysis, Critical Control Point (HACCP) is also a qualitative risk analysis method that has been included in Chap. 5 because of its original and continued association with food safety.

This is not an exhaustive list by any means, but it gives an idea of the types and applications of these useful and common methods of analyzing risks in almost any setting. Support for learning and developing qualitative techniques is well served by instructional facilities, traditional and online training programs, expertise, extension programs, and resources from both private and public sectors. Further readings and references are supplied at the end of the chapter.

4.1　Preliminary/Risk Analysis (PRA)

This is the simplest tool to use, perhaps at the beginning of an assessment when the available data are limited and the parameters are few. It can take two forms, a **linear** or descriptive form and a **tabular** form that can be created in a spreadsheet. PRA asks questions such as: *"What is the task?" "What could go wrong?" "What is the likelihood of these unwanted events?"*

4.1.1　PRA in Linear/Descriptive Format

This example involves transportation of a stock of vaccine that must be handled under "cold-chain" conditions, in which the temperature must not exceed −15 °C (−5 °F) at any time during transport or storage (Fig. 4.1). Travel usually takes 5 h and the insulated container is designed to retain −15 °C for 12 h.

4.1.2　PRA in Tabular Format

This version of the preliminary risk analysis allows the calculation of a simple *"risk index"* of seriousness or concern based on *probability × magnitude* that can assist in setting priorities and allocation of resources. For example, in Fig. 4.2, the local health authority is assessing the risk of dangerous pathogens being carried out from a biosafety level 4 (BL4) laboratory on hands or clothing of personnel.

PROJECT TITLE & DESCRIPTION:	Maintaining vaccine cold chain during 12 hr transit	
What will the process involve?	• Description of task/event/ process & rationale for analysis	·Transport of vaccine by air + road while maintaining cold-chain at −15°C (−5°F) in specially-designed, insulated containers
Hazard Identification	• What possible failures or faults? • Why/ how/ when/ where/ could each occur? • What factors could lead to each failures/faults?	Failure: temp exceeds −15°C, due to: ·Delay (exceeds 12 hr- capacity of container to maintain temp) or ·Container damaged en route
Risk Analysis/ characteri-zation	• How likely is a failure or fault to occur? (= "probability") • What would the consequences be? (= "magnitude") • Are there controls in place to detect/reduce these effects?	A) DELAY >12 hours uncommon, but subject to external uncertainties e.g., traffic, weather, accidents, diversions, vehicle breakdown. Prob: Moderate; Magnitude: (vacc. loss) serious B) DAMAGE: container is reliable. Prob: Low; Magnitude (vacc. loss) serious.
Risk reduction/ elimination	• What can be done to reduce the consequences, should an event occur?	·Known access to solid CO2 en route if needed, or transfer to alternative vehicle.
Risk Prevention & control	• What can be done to prevent a failure event? • How can we prepare for an event?	·Decrease uncertainty: (better planning, routes, equipment, contingencies) ·Back-up equipment, vehicles, training
Risk Monitoring and recording	• What factors/indicators should we be monitoring? • Where should we be monitoring? • When should we be monitoring?	·Constant-monitoring of temperature inside container (Min+Max) ·Use calibrated min/max data logger device with data download at start of transit, during, and arrival.
Risk Communi-cation	• At all stages, (planning, implementation, evaluation), does communication system ensure all stakeholders have prompt, effective, and clear information?	·Clarity, brevity, and immediacy in all communications, especially personal awareness and due diligence

Fig. 4.1 Example of preliminary risk analysis: vaccine cold chain

PROJECT TITLE & DESCRIPTION: Assessing risk of BL4 pathogens from facility.							
	1	2	3	4	5	6	7=>
Risk ID	Hazard, fault, or failure	Consequence	Contributing factors	Probability [P] of occurrence	[M] Magnitude /Severity	Risk Index	
	What could fail?	What would be expected to happen? Where? How?	What co-factors may lead to the failure?	Rating on scale 1_{low}–5_{high}	Rating on scale 1_{low}–5_{high}	Calc. PxM	**
A	Viable Biosecurity Level 4 pathogen escapes lab premises on workers' hands, clothing, bags, etc.	Pathogen infects worker within 72 hrs, and spreads to lab worker's family, community, school, etc.	Inadequate training of lab workers	1	4	4	**
B			Inadequate supervision of lab workers	2	3	6	**
C			Defective 'haz-mat' suits or glove boxes	3	3	9	**
D			Failed Safety System (protocol was bypassed)	4	4	16	**

** Column 7 would extend to preventive controls, options, monitoring-actions, etc.

Fig. 4.2 Preliminary risk analysis: tabular form

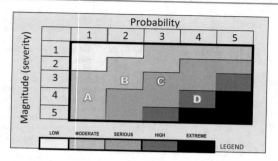

Fig. 4.3 "Heat map" depicting relative degrees of risk

The risk index is a product of values from two arbitrary scales (usually 1–5) and hence lacks any suggestion of external credibility but serves to prioritize the issues in a useful way.

The numbers are only relative. For example, [C] a defective hazard-materials suit (risk index = 9), is more immediately dangerous than [B] inadequate supervision (risk index 6), but presents less of a priority than [D] failing to comply with the established protocol/procedure, the "safety-system" (risk index 16).

The results can also be shown visually as a "heat map" (Fig. 4.3), where individual risk factors are placed on a grid with colors or cross-hatching identifying higher or lower priorities.

4.2   Failure Mode Effects Analysis (FMEA)

This technique can be applied to risks at any level of scale. The entire project or process can be analyzed, as can each individual component, right down to a switch, a thermocouple, or an O-ring. FMEA is one of the most widely used techniques, and it is usually completed using a spreadsheet. A further factor being considered is the degree to which the fault or failure can be detected. This is often omitted from simpler analyses, but experience with case studies from numerous fields and disciplines reveals that the recognition or identification of a fault or anomaly is often initially missed or mis-categorized, contributing to the overall degree of risk, as well as more severe consequences through delays or incorrect decisions (SAE 2018).

In the analysis shown in Fig. 4.4, the risk priority number (RPN) can be interpreted to indicate an appropriate response. An example of the interpretation is given in Fig. 4.5.

Clearly, risks 1b and 1d in Fig. 4.4, are considered "moderate" on this scale, but the risk represented by 1a is "major" and 1c is "significant," requiring specific actions.

FAILURE MODE EFFECTS ANALYSIS (FMEA)								
PROJECT:	Automatic chlorinator for rural hospital deep-well water supply							
RISK QUESTION:	What is the risk that the chlorinator will fail to provide safe water to the hospital?							
ID #	Process step	Failure mode	Failure effect	Failure mechanism	Likeli-hood L	Severity S	Detectability D	RPN
	Expected normal function	How could this fail?			1= low L 5= high L	1= low S 5= high S	1= easily det. 5= not easily detected.	LxSxD
1	Free chlorine level maintained at 2 ppm	No electr power	Unsafe water = risk of illness	At device	2	5	5	1a: =50
				Area blackout	2	5	3	1b: =30
		Hypo-chlorite supply depleted	Unsafe water = risk of Illness	Lack of monitoring & supply	3	5	4	1c: =60
		Low-chlorine alarm fails (if fitted)	Unsafe water = risk of illness	Sensor or alarm fails	1	5	5	1d: =25

Fig. 4.4 Failure Mode Effects Analysis: Hospital water supply example

Risk Priority Number (RPN)	Risk Category	Action/Response needed
90-125	Dangerous	Must be remedied immediately
60-89	Significant threat	Action needed: remediate at highest priority
40-59	Major threat	Action needed: Remediate by targeted completion date
18-39	Moderate threat	Should be addressed at first opportunity
1-17	Low threat	Remedial action required; do not ignore.

Fig. 4.5 Risk priority number interpretation

4.3   Root Cause Analysis (RCA) Using a Fault Tree (FTA)

The previous models were answering the question "What can go wrong, how likely, and what would be the consequences of such a failure?" Fault tree analysis (FTA) answers a somewhat different question: *"What are the factors and events that could lead to the failure being studied?"* As such, it is a "root cause analysis" (**RCA**) system used to identify the trail of events that could lead to, or contribute to, the "top event" that is the object of the analysis. FTA was originally developed in Bell laboratories for the US Air Force in 1962 but was later adopted by aerospace, automobile, chemical, nuclear, and software industries (OSHA 2016; US DHHS 2019).

FTA is a diagrammatic model that involves constructing a "tree" that is quite different to the probabilistic trees that were developed in Chap. 2. In FTA, the failure event or problem outcome is depicted at the top and is connected via *"gates"* to successive layers of factors or events that can directly or indirectly influence, contribute, or enhance the problem we are studying. The factors are depicted using standardized shapes or icons.

The rectangles are major issues/problems that need to be analyzed further. The diamonds indicate issues that could be developed further but are not the focus of the current analysis. The circles are "root causes" that we need to identify and act upon. The processes and faults are linked by means of *"and"* and *"or"* connectors or "gates." The *"and"* gate denotes that **all** the events must be present to cause the fault immediately above, while the *"or"* gate indicates that **any** of the events below may be sufficient to cause the fault above.

The FTA shown in Fig. 4.6 addresses the example of water supply in a rural hospital that we considered in the previous section. While the mechanical problems with the system remain to be taken up with the manufacturer, the root causes within the purview of the hospital (operator training and supervision) and the local authority (vulnerability to power outages) have been identified.

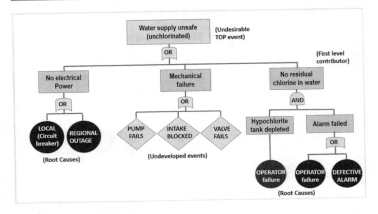

Fig. 4.6 Fault tree analysis

4.4 Management Oversight and Risk Tree (MORT) Analysis

The MORT procedure for analyzing causes and contributing factors began in the mid-1970s as a program to assist the US nuclear industry to achieve high standards of health and safety.

In the MORT context, accidents are defined as unplanned events that produce harm or damage (i.e., losses). Losses occur when a harmful agent comes into contact with a person or asset. Most of the effort in a MORT analysis is focused upon the identification of problems in the work or process and deficiencies in the protective barriers associated with it. The identified problems are then analyzed for their origins in planning or design stages, or in policy and procedures.

A MORT analysis first identifies the key steps in sequence. Each step is seen as (1) a vulnerable target (2) exposed to an agent of harm (3) in the absence of adequate barriers. The barrier analysis is key to the process, while the MORT chart acts as a "prompt" list allowing the analyst to focus on each of the issues revealed through the process. A color code is used as follows (Fig. 4.7):

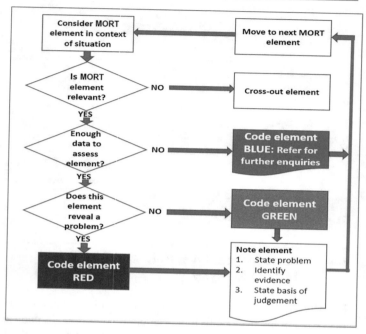

Fig. 4.7 MORT fault tree

- Red, where a problem is identified
- Green, where a relevant issue is judged to have been satisfactorily resolved
- Blue, to indicate a relevant issue lacking information to properly assess it

4.5 Hazard and Operability (HAZOP) Analysis

The disastrous release of 2,3,7,8-TCDD in Seveso (Italy) in 1976 and the Flixborough (UK) cyclohexane explosion in 1976 brought renewed focus upon the need for revisiting already approved plans and projects to review what components could become changed or altered, or how the specified processes and protocols could deviate from the original plan, thus introducing the risk of unforeseen disaster or catastrophe.

The HAZOP technique was initially developed in the 1960s as an aid to analyzing chemical manufacturing systems and batch processes, but it is now also used as the basis for reviewing a wider range of complex systems including refineries, nuclear power plant operation, decommissioning of reactors, and software systems and development.

A HAZOP study is a systematic investigation of a project or operation to identify and evaluate potential problems or deviations from the original design or engineering that may not have been considered initially and may not have been found using normal checklists and calculations. The overall process design is typically broken down into simpler sections or subprocesses called "nodes" that are examined individually.

The multidisciplinary HAZOP team is typically multidisciplinary, and the qualitative HAZOP technique stimulates the imagination of the experienced participants to identify potential hazards and problems in the operation. The approach is semi-structured in that the team applies standardized guide word "prompts" to each node being considered.

Each node, with its specified design intent, is commonly indicated on process flow diagrams (PFD), along with its complexity within the system and the magnitude of the potential hazards it might present. Identifying deviations is an iterative process based initially upon the application of a standardized set of guide words to each node being considered. When a deviation, real or potential, is identified, the team works back to identify the causes, typically by means of fault tree analysis (FTA). A typical set of guide words is shown in Fig. 4.8.

The words prompt discussion into the extent to which explicit parameters such as temperature, pressure, flow direction, composition, sequence, etc., could become affected and to what extent and consequence. The guide words are appropriate to the design parameter and are appropriately nonspecific.

For each deviation that is identified, the feasible causes and likely consequences are determined, and a decision is made (employing other risk analysis steps if necessary) as to whether

Guide Word	Meaning in relation to the intended design parameter
"No"	Absence or negation
"More" or "Less"	Increase or decrease
"Early" or "Late"	Variation to programed time
"Before" or After"	Variation to sequence
"As well as"	Unplanned addition
"Instead of"	Substitution
"Unequal"	Failure to balance (e.g., pressure)
"Reverse"	Wrong direction of flow
"Emergency shut-down"	Unforeseen sudden stop

Fig. 4.8 Example of HAZOP guide word list

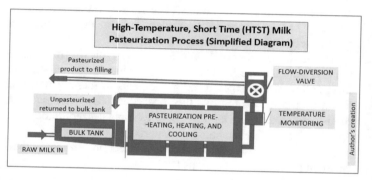

Fig. 4.9 Example of flow process for HAZOP analysis

the existing safeguards are sufficient, or whether additional safeguards are necessary to reduce the risks to an acceptable level.

In the continuous-flow HTST pasteurization process shown in Fig. 4.9, the flow-diversion valve (FDV) is a critical node. It is the means by which under-processed milk is immediately redirected to the bulk (input) tank and reprocessed. In the event that there is "*no*" diversion, or the diversion is "*late*" (HAZOP key words), the consequences would be under-processed (partly raw) product being distributed to consumers, with attendant health risks.

4.6   Case Study #12: Using FMEA and MORT

Often more than one tool is needed to determine the true root cause of the failure of an engineered system to meet expectations. An example of this was an investigation to determine the root cause of a fire in a natural gas heating stage for crude oil in a refinery facility. This analysis, made by a system safety engineer system safety engineer, describes the application of two approaches to risk assessment that have already been described earlier in this chapter.

Case Study #12: Risk Analysis as Part of the Investigation of an Oil Fire at a Refinery

During the refining process, crude oil is heated at various stages. This report describes events leading to an oil leak that was subsequently ignited by the natural gas heaters. The analysis of the event was made using both a Failure Mode Effects Analysis (FMEA) and a Management Oversight and Risk Tree (MORT) analysis.

The purpose of an FMEA (see also Sect. 4.2):

1. To describe the terminal event
2. To describe the failure modes leading to the event
3. To analyze the modes to determine the dominant mode
4. To verify the dominant mode through supporting data

The FMEA determined the following:

1. The dominant mode of failure was erosion-corrosion of the ferrous sulfide (FeS) layer inside the chromium steel elbows of heat exchanger tubes, caused by high-speed liquid droplets.
2. Naphthenic acid (NA), naturally in the crude oil, attacked the exposed steel leading to wall thinning of the elbows over a period of time.
3. The erosion-corrosion (1) and metal thinning (2) were caused by high velocity of fluid flow, high vaporization of fluid in the pipework, and/or inadequate corrosion resistance of the steel elbows.

4. The elbow components were designed to provide 15 years of working life when 5% chromium steel (P5) was used, given the specified total acid number (TAN) and sulfur content of the oil to be processed.
5. The P5 elbows began to leak at significantly less than the 15 years expected life of the heater tubes due to excessive wall thinning.
6. The P5 material was found inadequate for the type of crude oil that had been flowing through the heat exchanger.

At first examination, it would be natural to assume a design failure such as: *P5 steel appeared inadequate for the task and a more corrosion-resistant P9 steel should have been specified for the elbows.* While this certainly appeared to be true, there remained a need to get to the root cause and find out why the P5 steel became inadequate for the task. *Did the system and component designers make an error in specifying the materials? Or did the operating parameters subsequently change to the point where the P5 became inadequate?*

The MORT analysis (ICMA 2022) focused on the management/control system and the individual components that could lead directly to the failure. (See also Sect. 4.4)

The MORT analysis determined that:

1. The design process had been accurate and thorough. It was ensured that the materials used were adequate for the crude oil specified and for the lifetime expectation of the installation.
2. During the early deployment of the heat exchanger, the oil product being processed was exactly as specified in the design requirements. This included the TAN (total acid number) and the sulfur content. Therefore, the expected FeS corrosion-resistant layer buildup and the NA attack of this layer were as expected. Regular wall thickness measurements had been taken and no untow-

ard thinning was discovered. The heat exchanger was operating satisfactorily during initial operation.

3. Periodically, however, due to production demands, an operations management decision resulted in a different type of crude oil being introduced into the heat exchanger.

4. Not all crude oils have the same characteristics. At least one of the oils had a higher TAN and a lower sulfur content. While this different crude oil was in the heat exchanger, NA attack increased and the FeS protective layer decreased. The result was an increase in the rate of wall thinning. Upon returning to the original crude oil, the rate of wall thinning returned to normal, but the wall thickness had now reduced more than expected, as had the expected lifetime of the P5 elbow component.

5. The expected life of the components made from P5 alloy had been severely reduced by exposure outside the design specifications, that is, exposure to a crude oil with an increased TAN and decreased sulfur content.

6. Increased routine wall thickness inspections during heater maintenance would have shown that the wall thickness was reducing at an increased rate, prompting early replacement and preventing the leak and subsequent fire.

What can we learn from this analysis?

The FMEA identified that components had failed despite having been correctly designed for the purpose.

The MORT analysis determined that the system failure was caused by a management decision that overrode the design specifications. Although doubtless for good productivity reasons, the decision was made without adequate consideration about the impact on the life of critical components.

4.7   Case Study #13: Using HAZOP Analysis

The following analysis mentions several approaches to risk assessment that have been already described in this chapter, as well as considerations of uncertainty, and the danger of relying solely upon original ("as-intended") design formulae or calculations.

Case Study #13: Risk Analysis and Safety Considerations During Decommissioning of a Nuclear Reactor

When a machine has reached a point in its operational life where maintenance cannot guarantee safe performance cost-effectively, the machine has to be decommissioned. Decommissioning of aircraft, for instance, is a relatively simple dismantlement and recycling process, but nuclear reactors have very long lifetimes and present complex challenges, where the consequences of an accident can be severe.

In the decommissioning of equipment, the systems safety engineer has to answer the following questions:

1. *What condition is the equipment in that might impact safe dismantlement?*
2. *What activities are needed to ensure that dismantlement does not create a safety problem?*

The most common tools to answer these questions are:

- Probability Safety Analysis
- Deterministic Safety Analysis
- Failure Mode and Effect Analysis
- Fault tree analysis
- Management Oversight Risk Tree
- Hazard and operability study

In addition, there is always a level of uncertainty that must be considered when undertaking decommissioning activities where an unsafe condition may occur.

The SLOWPOKE reactor: In all, ten SLOWPOKE reactors were built. The *Safe Low Power Kritical Experimental* reactor is a low-energy, tank-in-pool research reactor designed by Atomic Energy of Canada Ltd. Two were low-power, 5 kW (thermal) reactors; the other eight were designated *SLOWPOKE 2*, with a power output of 20 kW (thermal) having a much greater sampling capability for experiments. Of these eight reactors, three are still operational and five have been shut down.

One of the reactors that was decommissioned between 2000 and 2004 had the following specifications:

The fuel was ceramic uranium dioxide with the uranium U-238 enriched (to 19.9%) with U-235. Despite being only 22 cm diameter and 23 cm in height, the core produces 20 kW (thermal) from 200 fuel pins. The fuel is suspended freely in a tank of light water (H_2O) enabling moderation, allowing fission, and providing fuel cooling.

The core is surrounded by a fixed beryllium metal annulus and a bottom beryllium slab. Beryllium acts as a moderator but also acts as a neutron reflector, reflecting neutrons back into the core. This process maintains a necessary configuration of neutrons in the core, to keep the reactor in a critical state. As the uranium fuel becomes consumed during operation, additional beryllium slabs are added to the top of the reactor to maintain criticality (Fig. 4.10).

During normal operation of the reactor, both the uranium core and the beryllium components become highly radioactive. Several isotopes in both materials produce ionizing radiation that is measured in half-lives of millions of years, so the amount of radiation shielding required during decommissioning is almost independent of the time that the reactor has been operational or has been shut down. The reactor and its reflectors continue to emit hazardous levels of ionizing radiation requiring shielding during dismantling and transportation.

A HAZOP analysis was chosen as the most appropriate process to begin with, because it would reveal hazards from

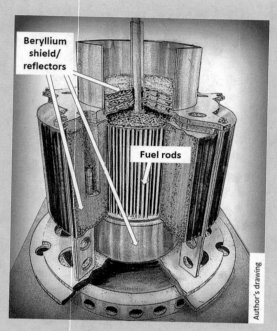

Fig. 4.10 SLOWPOKE reactor; drawing of cutaway model

processes, equipment, and system components and identify the additional tools required. The failure of any stage would have the potential to create unsafe conditions and/or significant delays in the safe removal of the core and all hazardous materials. (See also Sect. 4.5)

The initial analysis confirmed that the fuel and beryllium reflector-shielding calculations had been correct and that the quantity and type of shielding had been appropriate and correctly installed.

After the fuel had been safely removed from the reactor into a shielded flask, the beryllium reflector components were moved into shielded boxes. It was during this process that repeated measurements with portable instruments confirmed that one of the shielded boxes had a slightly higher external radiation signature than the others. Action was immediately taken to increase the shielding quantity, and then a root cause analysis (RCA) was conducted to determine the cause before work was allowed to continue. (See also Sect. 4.3)

While the drawings had indicated that the reactor had been correctly configured, this was found not to be exactly the case. The reactor core had not been suspended directly in the center of the pool and therefore not in the center of the beryllium reflectors. The reactor's performance had not been affected; it had operated safely for many years and had also been safely shut down for many years. But the beryllium reflector on one side of the reactor had been subjected to a higher level of irradiation than the other side, and therefore was slightly more radioactive, requiring an increased quantity of shielding during dismantling and transportation.

The anomaly had been detected immediately, so the higher external radiation emission did not impact the health and safety of workers because of other safety processes already in place to manage exposure time, distance, and shielding. The only delay to the project was the time needed to conduct the root cause analysis and confirm that the appropriate corrective actions had been taken.

What can we learn from this analysis?

Design specifications and calculations are never considered to be enough to guarantee safety. Invariably, there has to be more than one way of confirming that materials and processes are safe. Usually, this requires that accurate measurements are taken at every appropriate opportunity.

References

ICMA. International Crisis Management Association. Management oversight and risk tree (MORT). http://icma.org.uk/06-9_mort.html. Accessed Nov 2022.

OSHA. FactSheet: the importance of root cause analysis during incident investigation. 2016. https://www.osha.gov/sites/default/files/publications/OSHA3895.pdf. Accessed Dec 2022.

SAE. Fault/failure analysis procedure: design analysis procedure for failure modes, effects and criticality analysis (FMECA). Society for Automotive Engineers ARP926C. Feb 2018.

US DHHS. Root cause analysis. US Department of Health & Human Services. 7 Sept 2019. https://psnet.ahrq.gov/primer/root-cause-analysis. Accessed Dec 2022.

Risk Assessment in Food Safety and Foodborne Illness

5

Abstract

Foodborne diseases represent a major source of human morbidity and mortality. The World Health Organization's Foodborne Disease Burden Epidemiology Reference Group (WHO-FERG) estimated that in 2010 alone, foodborne diseases caused 600 million illnesses, 420,000 deaths, and 33 million Disability Adjusted Life Years (DALYs). Concern over increasing foodborne disease incidence has been expressed by governing bodies of the Food and Agriculture Organization (FAO), the Codex Alimentarius Commission, and food safety agencies around the world.

Whether the suggested increase is real or due to improved surveillance or more accurate detection of foodborne illness agents in foods is not clear. What *has* become clear in the last two decades is that the scale of international food production and distribution is such that even a small fault or error can sicken very large numbers of consumers. This chapter briefly introduces the methods currently available, and those being developed, to assess risks to food safety and prevention of foodborne illnesses. Further readings and resources are listed at the end of the chapter.

5.1   Scope of Foodborne Illness

Among common causes of acute ill-health, the foodborne ill-nesses group has been described as second only to the common cold in terms of incidence. Actual numbers of cases are not known, partly because most of this group are not notifiable or reportable, and partly because most occur as single cases or in small numbers, and the conditions remain largely undiagnosed and self-limited. Larger outbreaks, however, can involve hundreds or even thousands of people, and these are usually subject to intense investigation.[1] Foodborne disease outbreaks on this scale are frequently associated with listeriosis, non-typhoidal salmonellosis, or enterohemorrhagic *E. coli* infection (with hemolytic uremic syndrome as a complication). The economic and social impact of an investigation into a large outbreak can be considerable due to the illness (treatment, deaths), loss of productivity, food recalled or destroyed, disruption to businesses, loss of reputation, and from the investigation itself, personnel, laboratory work, and analysis.

Foodborne illnesses are caused by more than 200 known agents from bacterial, viral, parasitic, and fungal origins, as well as a range of toxicants and toxins. They grow in very different ways, in different foods at different rates, under a wide range of aerobic/anaerobic, redox, pH conditions, and water activities. Some are easily destroyed in seconds at 72 °C, while others survive many hours at 100 °C. The illnesses target different organ systems, at different rates, manifesting with symptoms ranging from gastro-enteritis to hemolysis, amnesia, renal failure, septic arthritis, paralysis, and sometimes death. Onset times vary between 20 min and 40 days. Attempts to control these agents and prevent food-

---

[1] The 1981 outbreak of a previously unknown condition was eventually traced to ingestion of an illegal cooking oil that had been adulterated with aniline dyes. It was later known as the Spanish toxic oil syndrome (TOS), and affected at least 20,000 people, of whom 300 died.

borne illnesses have evolved from antiquity through trial-and-error and avoiding mistakes of others, all the way to the astonishing advancements in food microbiology and molecular biology we have seen in the last 150 years.[2]

Methods to control enterohemorrhagic *Escherichia coli* in lettuce production do not resemble in any way the methods to control *Listeria monocytogenes* on cold cuts. Preventing the violent emesis and cramps arising from an encounter with enterotoxin-producing *Staphylococcus aureus* on sliced meats requires a different strategy to preventing paralysis and death from *Clostridium botulinum* neurotoxin in home-canned vegetables. Nevertheless, all these illnesses are essentially preventable, and from a systems perspective, these infections and intoxications can be seen as the result of several types of failure (Fig. 5.1).

System failures	Examples....
Failures of systems designed to control and prevent known causes of foodborne illness;	• Inadequate heating or processing, time-temperature abuse, incubation, cross-contamination events, especially between raw & cooked foods, dangerous food handling practices.
Failures of systems to anticipate internal or external changes in factors leading to outbreaks;	• Introduction of airtight wrapping on partially-dried fish, allowing anaerobic environment for C. botulinum growth; • Increased demand (and supply) of unpasteurized milk
Failures to anticipate new pathogens, or new 'paired' pathogen-food interactions.	• Emergence of Shiga-toxin producing enterohemorrhagic E. coli in ground meat, • Emergence of new aggressive strains of Vibrio parahaemolyticus in raw seafood, • Emergence of Cyclospora as contaminant in anti-fungal treatments of soft fruits.

Fig. 5.1 System failures contributing to foodborne diseases

---

[2] Prior to the early 1900s, most foodborne illnesses were simply described as "*ptomaine poisoning*" coming from the Greek "*ptoma*" = corpse. Anything decomposing was considered sufficient to cause these common ailments.

5.2   Root Cause Analysis (RCA) in General

If foodborne diseases can be justifiably described as failures on some level, it follows that unearthing and identifying the critical action(s), fault(s), default(s) omission(s), or error(s) should provide the means of understanding the problem and informing strategies designed to prevent it recurring again. Root cause analysis (RCA) has already been discussed in Chap. 4 in nonfood settings, but RCA methods have a logical place in food-related risk assessment. It is a systematic approach for identifying the underlying reasons (causes) why an event (e.g., an outbreak or food contamination) occurred. The goal is to reveal systemic weaknesses in the food system, such that the system can be adjusted or redesigned in a way that prevents repetition.

International, national, and local food safety agencies are primarily *preventive* in nature; thus, understanding the root causes of previous foodborne disease outbreaks or contamination events is essential for their work, as is the sharing of these results through publishing, presentation, and archiving. Errors have a way of repeating themselves, as inspectors and field investigators are well aware, and there is need to enhance dissemination of root cause analysis results (even with identifiers omitted) among food safety professionals, industry professionals, and personnel in regulatory agencies, as well as educators and academic researchers (Firestone et al. 2018).

5.3   Hazard Analysis, Critical Control Point (HACCP) Methods

5.3.1   HACCP Origins

HACCP was first developed in the early 1960s essentially as a food safety assurance protocol for the US space program. By the mid-1970s, HACCP principles were quickly implemented by large food manufacturers, possibly encouraged in part by several high-profile outbreaks of foodborne illness, and the system has since been adopted by a range of nonfood-related manufacturing, construction, engineering, institutional, and commercial sectors.

The HACCP system is designed to increase safety by means of careful identification of physical, chemical, and biological hazards, and establishing the means of controlling those hazards at every critical juncture. The scope begins with raw materials, their harvesting, processing, handling, and every step in manufacture, storage, distribution, and eventual consumption of the food item.

It is important for the HACCP assessor to be thoroughly familiar with the particular food production process, and with the relevant national and international standards and guidelines such as the Codex Alimentarius (2020).

Seven steps constitute an HACCP procedure:

1. Conduct a hazard analysis.
2. Determine critical control points (CCPs).
3. Establish critical limits.
4. Establish monitoring procedures.
5. Establish corrective actions.
6. Establish verification procedures.
7. Establish record-keeping and documentation procedures.

5.3.2 Identifying the Hazards

For a complex manufacturing system, an effective approach is to deploy a team of people with expertise from inside or outside the company that are familiar with the processing steps and especially the physical, chemical, and biological hazards that may be encountered throughout the process.

The hazards that could occur at each stage or process should be listed, including all ingredients, additives, packaging materials, cleaning materials, processes, and equipment.

- **Biological hazards** include bacteria, parasites, insects, rodents, birds, fungi, or viruses.
- **Physical hazards** include broken glass, metal staples, machine parts, or metal scrubbing bristles.
- **Chemical hazards** include cleaners, sanitizers, pesticides, and detergents. Special uses for the food should be kept in mind, such as infant formulae, baby food, or preparations for immune-compromised persons.

5.3.3   A Process Flow Diagram

A process flow diagram is an essential initial step in identifying the vulnerabilities or hazards and the key critical control points (CCPs). The diagram (Fig. 5.2) depicts an HACCP analysis on a bean-sprout processing plant (redrawn from template by CFIA 2014).

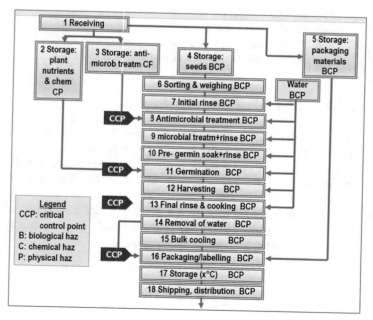

Fig. 5.2 Example of a process flow diagram for HACCP

The CCPs are essential actions or activities designed to intervene at a specified stage or point in the process, thus controlling, eliminating, or reducing the threat to permitted or acceptable levels. Omitting or bypassing a CCP represents a serious threat to the safety of the product, and adequate monitoring must be established to detect when a CCP has failed its specifications for whatever reason. Record-keeping and full documentation are essential.

5.3.4 Validation of the HACCP Plan

Validation of the HACCP plan is carried out to address two questions: (1) *Was the plan originally designed for this commodity manufactured by this specific production process, with full regard to the established technical principles and peer-reviewed scientific literature related to this production process?* (2) *Operationally, is the plan still being implemented exactly as it was designed?*

In addition, periodic (annual) audits are standard, as are reviews at any time when a new step is added, or a change made to the process, plant, or premises. Training of all staff has to be kept current, with new personnel undergoing instruction as soon as they come on board. Numerous excellent resources are available for further development of an HACCP plan and verification (FDA 2022; PHAC 2018).

5.3.5 Limitations of HACCP

While HACCP focuses on identifying where and when known risks are likely to occur in an established process, it lacks the ability to predict and describe threats and hazards that are unknown or unanticipated. For example, the discovery that chocolate novelties were the vehicle in an outbreak of Salmonellosis (80 cases in Canada and the USA) due to *Eastbourne* serotype during Christmas 1973 and again at Easter 1974 is an example of a novel vehicle for a known pathogen (Craven et al. 1975). The first isolation of Shiga toxin-producing enterohemorrhagic *E. coli* O157:H7, a hitherto unknown pathogen, in undercooked hamburger patties was an example of a novel food/pathogen "pair." It affected at least 47 people in Oregon and Michigan (Riley et al. 1983). It is doubtful that any HACCP system in place could have foreseen these events, much less prevented them.

5.3.6   The Difference Between HACCP and ISO 22000

Whereas the focus of HACCP is squarely upon food safety, the scope of the International Standards Organization (ISO) 22000 is wider, addressing the structure and process of food safety in the company. Certification to ISO 22000 is independent, meaning that an organization or industry can decide to pursue certification or not, but for companies planning to develop their international markets and activities, conversion of their HACCP certificate into a certificate based on ISO 22000 would seem advantageous (ISO 2018).

5.4   Microbiological Risk Assessment (MRA)

The majority of acute threats to food safety (aside from individual allergies, or sensitivities such as phenylketonuria) are due to the infective or toxigenic activities of microbiological agents. Assessing how effectively microbiological agents can influence risk of illness or can be controlled or eliminated is therefore central to reducing the risk of illness from food.

Sly and Ross (1982) used a 3×3 factorial design to confirm the correlation between selected microbiological parameters and certain hygiene indices in restaurants. For example, total aerobic plate count $>10^6$ was more strongly correlated with temperature control of foods ($r = 0.62$, $P < 0.05$), but poorly correlated with surface cleanliness ($r = 0.31$, $P > 0.05$). Though not technically a quantitative outcome, this approach was an early step toward measurement and prediction.

Poultry and poultry products remain a common source of salmonellosis. In setting priorities and practices to control the risk of salmonellosis for the consumer, that process of decision-making requires specific information about microbiological agents and the ways they can be controlled (Timoney et al., 1989). Risk assessment becomes an essential tool to inform and support this process. For example, *how much can risk of salmonellosis for the consumer be reduced if each of these actions are adopted?*

- Reduction of pathogen load by 90% in the gut of the birds in the broiler house by vaccination
- Reduction of pathogen load by 80% in the gut of the birds by competitive-exclusion techniques
- Reduction of *Salmonella* on the carcass through pressure washing
- Reduction of *Salmonella* on the carcass by dipping in hypochlorite rinse

To clarify, in the food microbiological setting, "hazard" applies to the source of the danger (e.g., the *Salmonella* pathogen), while the "risk" incorporates the probability of illness (recall: $risk = hazard \times probability$).

A microbiological risk assessment (MRA), particularly if it involves quantitative assessment, can be time-consuming and resource-intensive and requires the input of a multidisciplinary team. But the scale of the problem and its global scope justifies the effort to develop these techniques, especially in the anticipation of novel food sources, novel processing techniques, and attempts to feed a global population in the face of crop failures and reduction of traditional food supplies.

The Codex Alimentarius Commission (2014) has expressed a clear need for science-based tools to aid in decision-making about the microbiological safety of foods, with input from food scientists, food technologists, epidemiologists, microbiologists, and mathematical modellers at national and international levels.

An early example of a microbiological risk assessment in a specific pathogen-food pairing was published by the FAO/WHO working group (FAO/WHO 2002). It comprised two exposure models, each involving salmonellosis risk, one from eggs and the other in broiler chickens. Each model took the form:

| Hazard identification | => | Hazard characterization | => | Exposure assessment | => | Risk assessment |

The risk characterization for eggs estimated the likelihood of illness in a person ingesting a single serving of eggs either as whole eggs or in the form of a meal consisting partly of eggs. The analysis began at egg production, and considered both transovarian transmission and contaminated shell routes, as well as the processing of eggs into complex ingredients, handling at retail and consumer level, and preparation methods prior to consumption.

The risk of acquiring salmonellosis from consuming broiler chickens commenced following slaughter and considered all processing steps including in-home handling and cooking. The outcome could be expressed as either the individual risk per consumer of the food or the number of infections expected per million servings of these foods.

5.5   Quantitative Food Risk Assessment

5.5.1   Quantitative Risk Assessment-Epidemic Curve Prediction Model (QRA-EC)

The versatility and value of qualitative risk assessment systems such as HACCP and RCA are beyond doubt. But for some applications, a tool is needed that provides quantitative assessment. A quantitative microbiological risk assessment (QMRA) model for estimating adverse health outcomes from microbiological agents in food, water, and the environment was proposed by Hass et al. (1999). This model attempted to predict illness (I) having virulence (r), following exposure to a specific dose (D) of a pathogen, using $P(I \mid D) = 1-\exp(-rD)$.

Mokhtari et al. (2022) developed the "*Quantitative Risk Assessment-Epidemic Curve Prediction Model*" (QRA-EC) not only to analyze root causes but also to explore the possibilities of gaining insight into the epidemiological consequences of the identified faults. The method was deployed to evaluate the impact of several potential root causes identified during the 2019 outbreak of salmonellosis due to freshly cut melons (CDC 2019). The quantitative outcome was in the form of the number of predicted cases of salmonellosis that could be attributed to each of the faults or failures along the melon supply chain and to predict the timeline for appearance of these cases.

They found that the incidence of illness varied with the prevalence of *Salmonella* organisms on the external and internal surfaces of whole melons and sliced melons, and especially with the amount of contamination of the cutting and equipment surfaces. The timeline of cases was strongly associated with "equipment sanitation efficiency." This approach could be adapted to any pathogen-food pairs to identify the risk priorities and the potential incidence that could result from failures.

5.5.2 A Stochastic (Monte Carlo) Model for Multifactorial Analysis of Norovirus

Pouillot et al. (2022) devised a novel comprehensive quantitative risk assessment model that successfully evaluates the extent to which each of a number of specific factors influences the incidence of norovirus gastroenteritis due to the consumption of raw oysters. Their model explicitly evaluates the prevalence of viable virus at each of the steps in the pathway from a wastewater treatment facility on shore to the harvesting of raw oysters and their subsequent consumption. The model used stochastic methods to analyze a wide range of complex physical, biological, and environmental phenomena.

While the source of all norovirus (both genotype I and II) was the discharge of treated effluent from human sewage, a number of other factors were found to influence the predicted incidence of illness. Some of the most important of these (measured by the percentage increase in mean standard error) were the concentration of norovirus genotype II in the influent of the waste-water

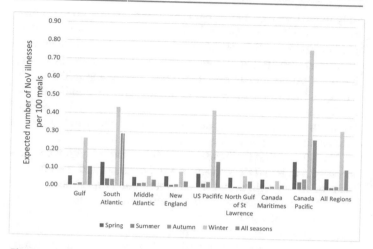

Fig. 5.3 Risk of NoV illness per 100 meals by location and season (Pouillot et al. 2022). (Copyright © 1999–2022 John Wiley & Sons, Inc. All rights reserved)

treatment plant (WWTP), the mean dilution of the virus in the estuary, the physical removal of virus in the WWTP, and the size of the meal ingested. By contrast, less influence was found from water temperature of the estuary, the time for the water to reach the harvest site, and WWTP ultraviolet disinfection procedures (Pouillot et al. 2022).

The model allowed the authors to predict an expected number of norovirus illnesses per 100 meals of raw oysters as 0.108 (USA) and 0.188 (Canada) when month and regional landings of oyster catch are integrated in the model (Fig. 5.3). National estimates for the annual burden of illness upon public health systems can be extrapolated, giving annual predictions of 60,255 cases (USA) and 6,860 cases (Canada), or 183 and 181 cases per million populations, respectfully, while assuming 50% of the oysters were actually consumed raw (Pouillot et al. 2022). The type of tertiary (final) treatment of the effluent was also found to be related to the consumption risk (Fig. 5.4).

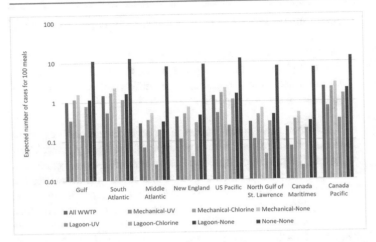

Fig. 5.4 Risk of NoV illness per 100 meals by location and effluent treatment (Pouillot et al. 2022). (Copyright © 1999–2022 John Wiley & Sons, Inc. All rights reserved)

The potential benefit to the public's health of an analysis such as this is obvious, and cannot be underestimated. As long as the consumption of raw oysters is permitted, specific factors can be measured, monitored, and controlled to reduce the risk of norovirus illness among raw-oyster consumers.

5.6   Canadian Food Inspection Agency ERA Model

The risk assessment model developed by the Canadian Food Inspection Agency (CFIA) was an attempt to allocate resources more effectively by focusing inspection and regulatory attention more specifically where it was needed most (CFIA 2021). Described as an "establishment-based risk assessment for domestic food establishments" or "ERA-Food," the model takes into consideration three different groups of risk factors:

- Inherent risk factors (e.g., food type, activities and processes, population served)
- Mitigation factors (e.g., food safety certifications, third party audits, QA personnel, sampling plans, audits of suppliers)
- Compliance factors (e.g., history of enforcement, recalls, complaints, inspection reports)

These factors, integrated through an adjustable algorithm, assist decisions about frequency and oversight of the agency's regulatory program. The program's baseline determination of the magnitude of potential hazard for food-pathogen combinations was determined in 2013, when 75 Canadian experts completed an online questionnaire assessing 155 risk factors and their potential impact on food safety. The program has been evaluated by determining how closely the ERA-Food model matched assessment by senior CFIA inspectors (CFIA 2015). The conclusion was that *"While substantial uncertainty around the central tendency estimates was found, these estimates provide a good basis for regulatory oversight and public health policy."* The complete findings of this expert group are published in Microbial Risk Analysis Journal (Zanabria et al. 2019).

References

Canadian Food Inspection Agency. Archived – HACCP generic model for sprouts grown in water. 2014. https://inspection.canada.ca/food-safety-for-industry/archived-food-guidance/safe-food-production-systems/haccp-generic-models-and-guidance-documents/generic-model-sprouts/eng/1368630463879/1368630464660. Accessed Dec 2022.

Canadian Food Inspection Agency. HACCP recognized establishments. Ottawa. Revised 2015. https://inspection.canada.ca/food-safety-for-industry/archived-food-guidance/safe-food-production-systems/food-safety-enhancement-program/recognized-establishments/eng/1299860323382/1299860380217. Accessed Nov 2022.

Canadian Food Inspection Agency. The Establishment-based Risk Assessment model for food establishments. Ottawa. Revised 2021. https://inspection.canada.ca/about-cfia/cfia-2025/era-models/era-model-for-food-establishments/eng/1551995065897/1551995066162. Accessed Feb 2023.

Centers for Disease Control (CDC). Outbreak of *Salmonella* infections linked to pre-cut melons. Final update. 24 May 2019. https://www.cdc.gov/salmonella/Carrau-04-19/index.html

Codex Alimentarius Commission. Principles and guidelines for the conduct of microbiological risk assessment. CAC/GL 30-1999. Food and Agriculture Organizations of the United Nations, and World Health Organization. Adopted 1999, Amendments 2012, 2014.

Codex Alimentarius Commission. General principles of food hygiene CXC 1-1969. Food and Agriculture Organizations of the United Nations, and World Health Organization; 2020.

Craven PC, Mackel DC, Baine WB, Barker WH, Gangarosa EJ. International outbreak of Salmonella Eastbourne infection traced to contaminated chocolate. Lancet. 1975;1(7910):788–92. https://doi.org/10.1016/s0140-6736(75)92446-0.

FAO/WHO. Risk assessments of Salmonella in eggs and broiler chickens Interpretative summary. World Health Organization Food and Agriculture Organization of the United Nations; 2002. https://www.fao.org/3/y4393e/y4393e.pdf

Firestone MJ, Hoelzer K, Hedberg C, Conroy CA, Guzewich JJ. Leveraging current opportunities to communicate lessons learned from root cause analysis to prevent foodborne illness outbreaks. Food Prot Trends. 2018;38(2):134–8. https://www.foodprotection.org/files/food-protection-trends/mar-apr-18-firestone.pdf

Hass CN, Rose JB, Gerba CP. Quantitative Microbial Risk Assessment. Wiley and Sones, New York. (1999);449.

International Standards Organization. ISO Food safety management 22000:2018. 2018. https://www.iso.org/iso-22000-food-safety-management.html. Accessed Oct 2011.

Mokhtari A, Pang H, Farakos SS, McKenna C, et al. Leveraging risk assessment for foodborne outbreak investigations: The Quantitative Risk Assessment-Epidemic Curve Prediction Model. Risk Anal. 2022;43(2):324–38. https://onlinelibrary.wiley.com/doi/10.1111/risa.13896

Pouillot R, Smith M, Van Doren JM, et al. Risk assessment of norovirus illness from consumption of raw oysters in the United States and in Canada. Risk Anal. 2022;42(2):344–69. https://doi.org/10.1111/risa.13755.

Public Health Agency of Canada (PHAC). Food safety. Updated 31 Jan 2018. https://www.canada.ca/en/public-health/services/food-safety.html. Accessed Nov 2022.

Riley LW, Remis RS, Helgerson SD, et al. Hemorrhagic colitis associated with a rare Escherichia coli serotype. N Engl J Med. 1983;308(12):681–5. https://doi.org/10.1056/NEJM198303243081203.

Sly T, Ross E. Relationship between hygiene and bacterial flora. J Food Prot. 1982;45(2):115–8.

Timoney JF, Shivaprasad HL, Baker RC, Rowe B. Egg transmission after infection of hens with Salmonella enteritidis phage type IV. Vet Rec. 1989;125:600–1.

U.S. Food & Drug Administration. Fish and fishery products hazards and controls guidance. 4th ed. Food and Drug Administration. Revised June 2022. https://www.fda.gov/food/seafood-guidance-documents-regulatory-information/fish-and-fishery-products-hazards-and-controls. Accessed Oct 2022.

WHO/FAO. Public Notice and comments on the second meeting of the WHO Foodborne Disease Burden Epidemiology Reference Group (FERG) 2021–2024. 7 Oct 2021. https://www.who.int/groups/foodborne-disease-burden-epidemiology-reference-group-(ferg)

Zanabria R, Racicot M, Leroux MA, et al. Source attribution at the food sub-product level for the development of the Canadian Food Inspection Agency risk assessment model. Int J Food Microbiol. 2019;305:108241.

Communicating Risk

Abstract

An occupational health and safety (OHS) poster urging workers to wear protective equipment, a road-sign warning of a dangerous intersection ahead, the health department urging mothers to get their children vaccinated, and a plan to prepare before the hurricane arrives are all examples of risk communication.

In this chapter, we leave the calculations behind and consider first how people perceive risk. Regardless of national statistics, does the worker consider that he or she is actually at risk from death by electrocution when servicing high-voltage equipment? Does the upwardly mobile urbanite even believe that unpasteurized milk obtained illegally is a risk to their children's health?

Building upon an understanding of risk perception, we can plan, design, and implement more effective means to share and communicate the quantitative information about risks that we have gathered.

The US National Academy of Sciences in 1989 defined risk communication as:

- An interactive process of exchange of information and opinion among individuals, groups, and institutions
    - Involving multiple messages about the nature of risk
    - Involving messages not strictly about risk, such as legal and institutional arrangements for risk management

The effective and successful communication of risk between all stakeholders in risk assessment activities takes on a larger and more central role than was assumed in previous decades. Numerous examples are on public record where ineffective communications have confused or angered the public, or delayed or even cancelled worthy projects.

6.1 We Have the Numbers, Now What?

Up to this point in the text, our objective has been to provide a hands-on introduction to the "numerical" aspects of risk assessment that can be usefully applied to most health or safety fields. These methods have been developed extensively in the last half-century, enhanced by improvements in toxicology, pharmacology, and molecular biology, and have resulted in analytical methods with validity, reliability, and reasonable precision.

We have calculated and characterized the risks; the job now is to share, communicate, and explain our results and be prepared to discuss them such that stakeholders, especially the citizen, can fully explore, digest, challenge, clarify, understand, and assimilate the information and make educated decisions and judgments about life, health, and safety (Fig. 6.1).

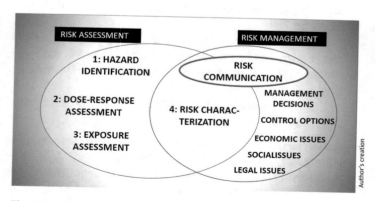

Fig. 6.1 Relationship between risk assessment and risk communication

Early attempts and even some recent examples of communicating risk to the public were not always "successful," sometimes constituting what Hance et al. (1989) described as "horror stories." Fishhoff (1995) pointed out that *"Getting the numbers right"* is only the first stage in a far more complex process involving at least eight steps:

1. *All we have to do is get the numbers right.*
2. *All we have to do is tell them the numbers.*
3. *All we have to do is explain what we mean by the numbers.*
4. *All we have to do is show them they've accepted similar risks in the past.*
5. *All we have to do is show them that it's a good deal for them.*
6. *All we have to do is treat them nice(ly).*
7. *All we have to do is make them partners.*
8. *All of the above.*

Most risk communication attempts, claims Fishhoff, barely accomplish the first two steps.

One of the goals of this chapter is to better understand how people *perceive* risk and the threats to their health and safety so that our communication plans can more accurately address those needs, fears, or perceptions. The goal of course is that communications should evolve to become steadily more effective, while mutual trust between the citizen, the "expert," and indeed all stakeholders might improve. These elements are deeply interwoven.

The need for critical review of the processes and methods that we use in communicating risks has never been more urgent. Electronic access to global databases and research libraries, made possible for the first time via the Internet, has resulted in a generally better-informed public, but at the same time, the gains we have made have been partially offset by the seemingly endless flow of readily available myth, superstition, misunderstanding, *dis*information, and conspiracies. The consequences are a confused public, hesitant about such obvious public health accomplishments as routine immunization of children, and some even willing to believe that vaccination is a tool of world domination.

Claude Shannon (1948) provided the basis for several decades of communication. That model has been one of the most common *obstacles* to effective and useful risk communication and is unfortunately still often seen today in both public and private sector communications (Fig. 6.2). The *"message"* originates with the *source* (the "expert"), via a selected *medium* (which today, in addition to radio, television, and print, includes many channels of social media), through transforming sources of *noise,* to the *receptor* (an awful euphemism for the citizen-layperson).

"Noise" is a valid dynamic here and includes all manner of opinion and influence, as well as social, cultural, fiscal, and political amplification or attenuation of the risk, and interactions with a sensitized or desensitized, aggressive, or even an apathetic public.

More importantly, this "one-way" approach allows no opportunity for feedback, much less dialogue, and presents an obstacle to modern society's need for collaboration and participation in decision-making. Fortunately, it is gradually being abandoned by modern communicators in favor of a more time-consuming, but ultimately more successful, *partnership* between the community and the "experts," or the "agency," whether from the private or public sector.

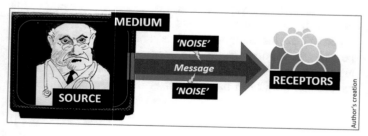

Fig. 6.2 *Obsolete model* of risk communication

6.2   The Decline in Trust and Credibility

6.2.1   Poor Management in High-Profile Health Crises

Between 1976 and 1996, several large-scale unfortunate events involving dioxins, PCBs, PPBs, pesticides, and some emerging diseases had been poorly managed by governments in the UK, the European countries, and the USA, leading to demands for greater transparency, accountability, and dialogue. Original incidence reports had been covered up, denied, withheld, or delayed. Requests by the media and the public for information typically remained unanswered, while requests for participation in the decision-making process were largely ignored or only superficially addressed. As a result, the public's long-standing trust in traditionally credible agencies, institutions, and authority figures began to wane, and the information, bulletins, and advice from these sources were given less credibility by the public and media.

One of the most noteworthy episodes in recent times centered on the mid-1980s outbreak in the UK of an unfamiliar disease of cattle. Fearful of damaging the reputation of UK beef and dairy markets and exports, the Ministry of Agriculture, Fisheries, and Foods (MAFF) asserted for 10 years that bovine spongiform encephalopathy (BSE) was only an animal disease and that British beef was safe. On March 21, 1996, in a spectacular about-turn, the MAFF announced that the first ten human cases were being investigated of a disease that appeared to be BSE transmitted to people. The loss of credibility for MAFF was profound. Further details of this and other misjudgments can be found in selected Communication Case Studies (Sect. 6.11).

6.2.2   Loss of Trust in Traditional Sources

This had been described in the USA by Kasperson (1988) and others, while in the UK, even slightly *before* the BSE *volte-face* in 1996, Frewer and colleagues at Reading University were busy

Trust, Information, and Food-Related Hazards

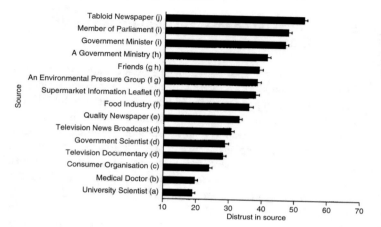

Fig. 6.3 Relative sources of "distrust". (Used with permission of John Wiley & Sons, Inc., from Frewer et al. (1996). © Society for Risk Analysis. Permission conveyed through Copyright Clearance Center, Inc.)

investigating which sources of information were "*most distrusted*" by the public. Central government sources were found near the top of the list (Fig. 6.3). If this survey were to be repeated today, we wonder where social media "silos" and "echo chambers" would be positioned, or the increasingly partisan private broadcasters and bloggers, unchallenged by the decreasing consumption of more "balanced" traditional print and broadcast media (Löfstedt, (2013).

Respondents were asked: "*Each of the following sources has provided information about*" … "[– a series of risks including alcohol use, food irradiation, food poisoning, genetic engineering, natural toxins, pesticide residues etc.]". These results (Fig. 6.3) show only the respondents' *distrust*. (Where parenthetical letters differ, statistical significance $P < 0.05$ was found).

In the USA, the Pew Research Center commented in 2010:

"*…By almost every conceivable measure, Americans are less positive and more critical of government these days. A new Pew Research Center survey finds a perfect storm of conditions associated with distrust of government – a dismal economy, an unhappy* *public, bitter partisan-based backlash, and epic discontent with Congress and elected officials..."*

The recent Covid-19 pandemic revealed innumerable instances where authentic "expert" sources were questioned and even ridiculed on an unprecedented scale. While questioning is not in itself unhealthy and, indeed, is key to scientific inquiry through challenge and hypothesis testing, we see countless examples of valid and verified information and advice being purposefully rejected and ignored in favor of anti-science misinformation, conspiracy theories, false commentary, and unfounded opinion from TV celebrities and kitchen-table websites.

6.2.3   A Major Obstacle to Effective Communication

Peter Sandman (2012b) had underscored very firmly that trust and credibility determine not only the degree to which the communicator was believed, but they also determine the extent to which the information and the message are accepted. He noted:

> *"The most important finding ... The agency's behavior, and the agency-community relationship have a substantial impact on the public's perception of risk ... more impact than the objective seriousness of the risk, and far more impact than any technical explanation of the risk".* (Sandman 2012b)

Trust cannot be bought, transferred, acquired, or generated in the short term. An individual or an agency must *earn trust* through performance, consistency, and reliability. If you are "parachuted" into a community who don't know you, you cannot expect them to trust you initially.

6.2.4   A Proxy for Trust

Perhaps the closest "proxy" for trustworthiness in the short term is *accountability*. While your *bona fides* will slowly become established, your status can be strengthened through addressing accountability, transparency, reliability, responsiveness, and consistency. Make sure you follow through with any requested information or inquiries left on answering machines; ensure questions are answered in understandable terms, with further opportunities offered for clarification. Whether arranging an information-exchange session or a town hall meeting, ensure that all arrangements are made in a transparent way, with local stakeholders having input into the organization, agenda, and logistics – even if you are simply arranging who provides the refreshments.

6.2.5 The Risk Information "Gap"

In any exchange of complex information and communication about risk, gaps and discrepancies can be expected to arise from differences in perception and awareness between the "expert" source of the information and the receivers of that information (Fig. 6.4).

This "space" or "gap" and its uncertainties were addressed by Powell and Leiss in 1997, who described this intermediate zone as a *"risk information vacuum."* Society and nature, the authors argue, abhor a vacuum, with the result that all manner of myth, speculation, and imagined fears quickly populate the space, creating a time-wasting and expensive distraction. And we can speculate that if a third party with specific interests and agenda manages to fill the vacuum with biased or prejudiced content, the impediment becomes a ponderous obstacle, guaranteed to delay or even prevent forward progress.

This underscores again the essential nature of transparency, clarity, and unambiguousness in all risk communication activities.

Fig. 6.4 Gap between experts and the public

6.3 Layperson Versus Expert: Two Perceptions of Risk

The expert's relatively uncomplicated view of "risk" as the product of *probability* and *magnitude* is not typically shared by the general public who employ a surprisingly complex range of criteria by which an issue, threat, product, or event is judged dangerous and best avoided. This phenomenon was observed and studied by Slovic (1987), Sandman (1983, 2012b), Hance et al. (1988), and others.

Hance et al. (1988) and Sandman (2012b) coined the term for these mostly qualitative criteria as "*outrage*" factors. Although lacking in true objective validity, these heuristic devices should not be disregarded as artifacts of poor education or misunderstanding (Sandman 2012b). They are measurable, predictable, and controllable and have very real effects.

Although more than 60 issues have been identified (e.g., Covello 1995; Hance et al. 1988; Sandman 2012b), a shorter selection is shown here relating more closely to health and safety fields.[1]

1. **Is it voluntary or coerced?**
2. **Is it natural or industrial?**
3. **Is it familiar or exotic?**
4. **Is it not memorable or memorable?**
5. **Is it not dreaded or dreaded?**
6. **Is it chronic or catastrophic?**
7. **Is it well known or not well known?**
8. **Is it controllable by me or by others?**
9. **Is it considered fair or unfair?**
10. **Is it morally irrelevant or morally relevant?**
11. **Is the source trusted or not trusted?**
12. **Is the process responsive or unresponsive?**
13. **Are non-vulnerable or vulnerable people affected?**
14. **Are the effects immediate or delayed?**
15. **Is it harmless or harmful to kids or unborn?**
16. **Are victims non-identifiable or identifiable?**
17. **Is media attention low or high?**

[1]These 17 factors were selected from 20 listed by Covello and Sandman (2001).

Each "outrage" factor is shown as a pair or couplet; if the topic or subject of the message tends toward the left side (e.g., "*natural*"), an easier passage is predicted for public consideration. A characteristic tending more to the right (e.g., "*industrial*") heralds a bumpy ride, with increased resistance, argument, and conflict, often with no clear relationship to empirical evidence or objective analysis. We have already discussed (#11) the importance of trust in risk perception in Sect. 6.2.

6.4   Understanding the Dynamics of Outrage

These "outrage factors" are nearly universal across cultures, traditions, age groups, and socioeconomic backgrounds. They are more complex in scope than the simple $R = M \times P$ definition of risk, and they are also to some extent, predictable.

It is worth remembering, too, that "experts" are also members of the public, and when they are not in the public spotlight, they, too, can be found responding to these heuristics. The Society of Risk Analysis conference was held in August 1996 at the University of Sussex, UK, and experts in risk assessment from many countries were assembled for the opening session. At the end of the evening, all the cheese, wine, canapés, and grapes had been eaten, but a large plate of roast beef sandwiches remained completely untouched. Five months earlier, bovine spongiform encephalopathy (BSE) transmission to humans had been confirmed, and while prime roast beef from young animals was not implicated, these delegates all preferred to avoid it entirely. The outrage factors in this example probably included those listed as #1, 2, 3, 4, 5, 7, 8, 11, 12, 14, 16, and 17.

Sandman and colleagues further proposed (2012a) that the clear distinction in *perception* of risk between the "expert" and the "lay" person can be traced to differences in the *definition* of risk. In assessing the same issue, the expert places almost all significance upon the objective probability × magnitude, whereas the public tends to weigh the *outrage factors* more heavily. When the *real* (objective) risk and the *perceived* risk (outrage) are both high or both low, not much opportunity for discord or argument exists. For example, a community feels threatened and anxious by several cases of meningococcal disease at a local school (high perceived risk: factors: 1, 3, 5, 8, 15, 17), and the authorities respond quickly with all the requested information and appropriate advice in understandable phraseology, as well as the precautions and safeguards the community needs. There is little or no conflict, and the communication is considered effective.

But increased controversy, arguments, and delays are more likely when objective risk and perceived risk/outrage are in discord. For example, the community is convinced that the highly visible white vapor emanating daily from a petrochemical refinery smoke stack represents a danger to the families in the community (i.e., high outrage: 1, 2, 3, 4, 7, 8, 14, 15). The health department, on the other hand, have environmental monitoring data showing the vapor to be just harmless steam condensate (low risk). But if the complaints of the community are simply refuted by the health department, with implications that the anxiety is due to imagination and lack of knowledge, the suspicion about the agency increases, trust decreases, and #11 and 12 are quickly added to the outrage list. This rift could have been prevented by inviting several community representatives to participate in a fully transparent monitoring, analysis, and reporting process, with full opportunity to inquire and discuss.

A municipal engineer's department is concerned about permitting houses to be built on designated floodplain land. Developers and prospective buyers, on the other hand, are determined to complete the construction and sale of "dream homes," because they cannot recall ever having seen a flood! (Low outrage, high risk). The discrepancy may be reduced by opening up the decision-making process, including updated predictions for climate extremes and detailed examples of severe losses in similar communities where planning restrictions had been relaxed or ignored.

Recent pandemic scenarios (2020–2022) have provided spectacular examples of high-risk-low-outrage conflict. Public health personnel encourage essential protection against viral transmission in high-density locations supported by global evidence and science (high risk) but encounter groups of community members who may have decided that there is low or no risk.

Experience has shown that ignoring outrage factors can be costly and disruptive. Failure to properly acknowledge and address perceived risk with the same resources and sincerity as the objective risk invariably causes delays, losses in finance, effort, opportunity, and on occasion cancellation of the project. One of the authors studied the perception of risk among residents living close to the site of a proposed large landfill facility in Ontario. Although the design was state of the art in technology and safeguards, the facility had been delayed for 11 years due to strongly held beliefs and negative perceptions. The expected life of the facility, however, once it opened, was only 10 years (Sly 1997).

Some of the groundbreaking work in risk perception was begun by Chauncey Starr, who found that we accept approximately three orders of magnitude (a thousand times) more risk when the risk is undertaken voluntarily than when we are coerced (item #1 in Fig. 6.6). A person attends a town hall meeting, arguing vehemently that a 10^{-6} lifetime carcinogenic risk from trihalomethanes in the town's drinking water is completely unacceptable. But during the break in the proceedings, the same individual may be seen calmly smoking a cigarette, a habit which carries closer to a 10^{-2} lifetime risk of death due to cancer. Clearly, the exposure to trihalomethanes were involuntary, but the smoking was a personal choice (Starr 1969).

As Starr describes it, *"We are loath to let others do unto us what we happily do to ourselves"* (Diamond 2012).

6.5 Identifying Different Needs, Roles, and Approaches to Risk Communication

6.5.1 Four Communication Models

We identify at least four distinct scenarios or roles for risk communication, each involving their own characteristics, motivation, and conflicts and requiring their own approach (Fig. 6.5).

Community motivation	Examples	Conflict expected	Perception of risk
Type 1: Need served In a crisis, the community wants the information, and it is provided.	• Advice about how long food stays cold in a power outage • A bulletin to expect storms and flash flooding in the next 48 hours	Conflict is low or absent	Both the *perceived* risk and the *actual* risk are high.
Type 2: Shake up Community members are apathetic or don't believe the danger exists or don't perceive themselves to be at risk.	• Early anti-smoking programs • First attempts encouraging seat-belt use in cars • Encouragement for mask use and vaccination in a pandemic where certain groups do not believe they are at risk	*Moderate conflict.* Danger may be clearly imminent, but many in the public deny or refuse to acknowledge the risk, and resist or refuse the necessary precautions.	The *perceived* risk is low, while the *actual* risk is high.
Type 3: Calm down Community members *believe* their health or safety is at risk to a far greater extent than the evidence can support.	• A trace pollutant in ground water • A preservative in food • A plasticizer in packaging • A belief (false) that the vaccine *causes* the illness it was designed to prevent	**Can be highly contentious,** with emotions and motivation extending to public protests and civil disobedience.	The *perceived* risk is far higher than the real risk, and *anxiety* emerges as the real hazard!
Type 4: False path Community members are convinced alternative remedies, actions, or solutions are more effective than those supported by evidence.	• Hydroxychloroquine or other ineffective 'remedies' for Covid19 • False advice given to parents about useless treatments for children with leukaemia • Beliefs that prayer, vitamins, or 'superfoods' will 'cure' cancer	If an alternative is dangerous or selected at the expense of an effective solution, **disagreement can be heated and may involve ethical arguments and legal intervention.**	Strongly-held (false) beliefs and perceptions are vigorously defended while rejecting evidence and experience.

Fig. 6.5 Four major types of risk communication

Type 1 communications should be straightforward: information is being sought, and it is provided, whereas communicating in types 2, 3, and 4, we expect some disagreement.

Apathy (type 2) can be long-standing, and the community may simply not believe that building housing on floodplain land or areas subject to erosion will affect them in their lifetime. Unfortunately, 100-year storms are becoming more frequent.

In #3 and #4, some members of the public can quickly become aggressive because they believe perhaps that *the agency is not levelling with us*, or is *covering something up*. In the last decade, and especially throughout the Covid-19 pandemic, pervasive social media elements have amplified this discord, fueled by the proliferation of unedited, unverified, unchecked, and misdirected "information" sources, effectively drowning out the invariably subdued official voices of fact-checked reality and empirical evidence.

Under such circumstances, trust and credibility are inevitably among the early casualties for *all* stakeholders. The community is frustrated by the agency's denial of what they see as a threat, while the agency, frustrated by unsuccessful attempts at communication, begins to doubt the ability of the public to understand the complexities of the issue – a dangerous assumption. It is always to be remembered that the majority of the community are generally aligned with the agency's advice but remain silent participants.

6.5.2   Anticipating Conflict and Disagreement

Communications involving potential conflict may have to resort to a slower, more nuanced approach during which extra time is taken to explore and acknowledge preexisting individual beliefs and myths and to counter misinformation. The language, terminology, and the methods used are crucial to establishing and maintaining trust in the agency and the individuals involved in the communication. Without that, the "message" itself will lack credibility. We have recently seen, for example, crowds outside hospitals denying the incidence and severity of the illness suffered by Covid-19 cases and denying the stress being experienced by overloaded intensive care units (ICUs). Proximity to the events and objective reporting by responsible authorities through neutral media appear not to sway the conviction of the misinformed zealot in the group setting.

Unfortunately, numerous examples of botched or delayed or poorly planned communications on the international stage remain in the collective memory of the public and the media, sowing the seeds for further distrust.

Ultimately, if education and dialogue fail, incentives and even legislation and enforcement may be necessary. For example, seat belts in cars were rare before the mid-1970s, but because cars were such a familiar part of our existence, protecting oneself against impact was not seen as a priority, and it took legislation requiring manufacturers to install belts as standard equipment and to require users to "buckle up" under threat of a fine, before general compliance was reached. The building of dwellings on floodplain land may be prevented by policy and bylaw, but this is frequently challenged by prospective homeowners and land developers, who petition for exceptions, despite the threat of catastrophic inundation. At times, the municipality may even grant the exemption because no one in the office can recall the last time the river flooded – an illustration of the "*NIMTOF*" principle ("*not in my term of office*").

An excellent comprehensive description of risk communication models and approaches from a wide range of practical and field contexts can be found in the highly recommended *Risk Communication: A Handbook for Communicating Environmental, Safety, and Health Risks*, 6th ed. by Regina Lundgren and Andrea McMakin (2018).

6.6 Inherent Difficulties to Be Prepared for in All Risk Communication

In addition to the "negative" characteristic inherent in each of the outrage factor couplets, a number of other phenomena can impede the progress or acceptance of a risk communication message.

6.6.1 Asymmetry Produced by Media in Attempting "Balance"

With noble intent, media agencies often insist on presenting *both sides of a contentious debate*. The problem is that while the two representatives are asked to present their differing viewpoints, one might be speaking for 99.5% of scientists in that field, whereas the other represents only 0.5% of current thinking. The one-to-one televised debate, however, gives the false impression that the two opposing views are roughly equal in validity and popularity. This perception could conceivably influence public opinion away from reality and toward the fallacy.

6.6.2 "Dueling PhDs"[2]

The public discussion of a technical nature may not be of great substance to the outcome, but spectators can become so disillusioned by the disagreement of highly credentialed individuals that they think, "If these people cannot agree, perhaps no one really knows!" Detailed technical debates may be best resolved before sharing publicly, although for transparency, care must be taken to follow through in full with the with now-clarified position.

6.6.3 Changes in Estimates or New Information

Public confidence is strongest when accepted information and explanations are consistent, and survive the storms of controversy unchanged. Science, however, must be sensitive to and reassess its conclusions on the basis of new evidence, discoveries, and analyses. This inevitably leads to doubts: *"Why are they changing their minds? How does the story keep changing?"* Scientists should take every opportunity to explain that as new information is found, it is assessed and incorporated into the message. This is also an opportunity to specifically announce and explain the arrival of the new information (perhaps the publication of a new piece of research) and to demonstrate and interpret the ways in which it adjusts the understanding of the issue. This would avoid the appearance of an "expert" suddenly announcing a different message but with no clear rationale.

---

[2]This phrase was quoted by Peter Sandman and he attributed it to Lois Gibbs, who established such excellent leadership in the Love Canal environmental disaster, New York.

6.6.4 Assurances That Are Too Assertive

A message that insists that there is "absolutely no risk," or that a procedure, water supply, bridge, pipeline, etc. is "100% safe," not only sounds too good to be true; it is often found to be flawed. On the other hand, admitting uncertainty or pointing out the incompleteness of the information can often *increase* trust in the source and, by association, the credibility of the information itself. An illustration of this phenomenon is presented as **Communication Case Study #15** at the end of this chapter.

6.6.5 Be Alert to Sensitizing Events

Be alert to sensitizing events from news reports, documentaries, or even recent movies. The announcement of the opening of a (hitherto unannounced) biosafety level 4 laboratory in Etobicoke, Ontario, was made just 3 months after the release of the Hollywood film *Outbreak*, in which a deadly virus resembling Ebola was seen to spread via a global pandemic. Neighborhood resistance was very strong, and the laboratory was ultimately not used for BL-4 work.

6.6.6 The Untrusted Messenger

The message or information will only be trusted as far as the messenger is trusted. If you are that messenger, and they've never seen you before, make an effort to link publicly with trusted, credible individuals in the community. In the meantime, your every statement, action, promise, and follow-through must be flawless and transparent. This will begin to establish *accountability*, a short-term substitute for *trust* that is acquired very slowly, and always by actions, never by the spoken word. For a more detailed discussion of trust, see Sect. 6.2.

6.6.7   Language and Meaning in Risk Assessment: The "Conservative Estimate"

Risk estimates are typically made based upon the worst-case (highest-risk) estimates of all variables and calculated using the upper bound (often the upper 95% confidence interval) of the slope factor. This is described as a "conservative" estimate on the principle that almost all the public would be exposed to a risk that was *less* than this. The implication is that safeguards and protections implemented to protect the maximally exposed individual (MEI) would automatically protect everyone else. This is completely opposite to the popular concept of a "*conservative estimate*," which is deliberately on the "low" side to avoid overestimating. This discrepancy may easily cause serious misunderstanding, and communicators would be well advised to suspect it and clarify for the public's understanding when detected.

6.6.8   Clarity and Transparency

An agency that uses jargon and technical "bafflegab" obfuscates the message and alienates the public. "*No comment*" responses, or any apparent reluctance to discuss, respond, and explain, will create more distance. Such miscommunications create what Powell and Leiss (1997) described as a "*Risk Communication Vacuum*," which rapidly becomes the focus of imagined cover-ups and conspiracies. Even when the lid is finally removed to reveal its non-conspiratorial contents, much effort is wasted to regain trust. All activities, even if only deciding the caterer for the meeting, should be transparent and open.

Simple, effective, language is also needed by many of the group you wish to reach for whom English may not be their first language.

6.6.9   Very Large or Very Small Numbers

Unless a person is familiar with numbers containing a lot of zeros, either to the left or the right of the decimal point, they tend to lose objective meaning. In a discussion of a precise risk estimate, you may hear someone counter with: "*one in a ten thousand – or one in ten million – it's all the same!*" Clearly, there are three orders of magnitude difference, but very large and very small numbers can seem beyond the cognitive grasp of many people. Try to keep the illustrations and comparisons to no more than "thousands" or "thousandths," because numbers in this range are more familiar in everyday life. (For a good illustration, see **Communication Case Study #14** at the end of this chapter.)

6.6.10 Exponential Misconceptions

The human brain is not naturally wired to understand exponents, and this characteristic is often exploited by illusionists and those keen on posing number problems. Since so many risk assessments result in exponents, we need to exercise caution in relaying such values while trying to ensure that they have been realistically understood. An extreme example of this inherent human weakness is embodied in the following scenario: Consider a power station with a 10-year expected life. For the first year (because the structure and staff are new), the risk of catastrophic failure is 1 in a hundred (or 0.01). For each of the next 8 years, the annual risk is 1 in 10,000 (or 0.0001), and for the last year, the risk is 1 in a 1,000,000 (that's 0.000001). There are no obvious exponents used, but ask about the cumulative risk for the entire 10-year life, and a good proportion of the guesses will be around 8 in 10,000 on the basis that the highest (1 in a hundred) and lowest (1 in a million) somehow cancel each other leaving something close to the middle 8 years times eight. The actual answer is 0.010801, or 1 in 92.6 for the entire 10 years, just slightly more than 1 in 100, which was the risk for year 1.

6.6.11 Comparing Risks for Better Effect

Other than on a spreadsheet, most risks don't exist in isolation. A process or activity may be assessed as carrying a certain risk, but so also will many alternative processes or activities used to achieve the same result. Stakeholders will benefit greatly by having a *"risk-risk comparison"* between a selection of parallel processes or activities.

But great care should be taken to ensure that comparisons are really made between equivalent alternatives in a similar context. For example, a community's groundwater contaminated with trihalomethanes (THM) from a nearby electronics plant may be relatively low on the risk scale, but a statement that claims the risk is lower than the risk from two glasses of wine every week misses the point. The intake of wine is voluntary and an individual's choice, taken in full awareness of the associated dangers, but the THM in water was not wanted, not requested, imposed by industry, and completely unsuspected until a chance water sample was taken. On no criterion is this comparison valid or acceptable.

Instead, compare risks presented by the activity with risks from similar or equivalent activities, such as all the available optional technologies for water treatment of a housing development.

Often, the greatest comparison or contrast can be found between the studied (incremental) risk and the background risk. This is particularly true, for example, with any discussion of cancer risk. All the conditions carrying a "cancer" label carry a connotation of dread, and yet between 1 in 4 and 1 in 5 of us will succumb to it. While heated debates and anxiety levels escalate over the additional lifetime fatality risk from a contaminant in water of perhaps 1 in a million, the community members who are holding forth, threatening civil disobedience, are already at a probability of lifetime cancer death of about 230,000 in a million. The focus of their anxiety raises their risk to 230,001 in a million. That incremental risk has not changed, but it can be revealed in a more realistic way by comparing to the background risk (see Sect. 1.5.1, and especially Communication Case Study #14).

6.7   **The Media Interview**

6.7.1   **Cultivate Relationships with the Media**

Cultivate relationships with the media – all types of print and broadcast. They can be a real asset to you in meeting your communicative objectives. A mutually respectful association with the media is an advantage to all parties; the reporters and journalists know they can always get an informed opinion, with facts explained for easy consumption, devoid of bafflegab or technical jargon. As a subject-resource person, you will have developed a route for getting your message out to the public with minimal distortion.

If a reporter or journalist asks for an interview, respond as soon as you can, whether to agree or to decline. Reporters have tight deadlines and will appreciate the promptness of the response. If you cannot add anything useful, or the issue is outside your field of expertise, explain promptly, and if you can, suggest one or two colleagues who may be able to assist.

6.7.2   **Preparing for the Interview**

Be clear about your PCO (*primary communication objectives*). This means writing down and memorizing the key points you wish to get across in the interview. Your job will be to bring your responses around to those key points. Don't totally evade the original question, but try to include the key points in the answer.

Remember also that while you are the *subject expert*, the reporter is a trained and experienced expert in seeking out information and identifying a story. Try to avoid giving distorted or vague facts; search carefully through your own words, the phrases you're using, and the details you're giving. Are you rushing through part of the explanation, or should you have gone into greater detail in simple language to make sure the details were clear? It doesn't hurt to ask after the interview if your responses were clear, and whether you can clarify anything. Topics and terminology that are familiar to you on a daily basis may be new or unfamiliar to your interviewer and to the public.

6.7.3   The Media Want an Interview *Now*

If it's **not** an immediate crisis, and you don't have the information you need, explain this and reschedule for the afternoon or next day. Tell the news crew that their viewers or listeners deserve to have full and complete information and that you will be pleased to do an interview, say, at 9 am tomorrow. The media people will object, BUT they *will* be there at 9 in the morning. In the meantime, bring in or talk to your field people. You learn exactly what has happened up to the present time, numbers, details, quantities, locations, and next steps. Together with your staff, you go through responses to the questions you are likely to be given, to make sure they are current, consistent, accurate, and precise, including that one "awkward" question. *THAT will probably be the first question they ask!* Remember they are professionals at digging out the leads to a story. The mistake would have been to agree to an interview unprepared and uninformed, which would have turned the story away from the original topic, and refocus into: "*…why is this person in charge of this important issue when they clearly don't know what is happening?*" (For an in-depth example, see Communication Case Study #17.)

However, if the matter *is an urgent health or safety crisis*, an interview should be given but if the information is uncertain, unreliable, or incomplete, *that fact must be made first, very clearly* at the outset, with promises to keep the media updated without delay as the new information becomes available. The precautionary principle is paramount. Do not wait until you have all the facts and they have been certified 100% accurate. In an *urgent* situation, share what you have in full detail (see Communication Case Studies #18, 19, 20 and Sect. 6.10).

6.7.4 The Interview: Checklist

- **"Off the record."** The rule is that when there's a reporter in the room, or a microphone, or camera, you should not rely on anything being "off the record." (Case Study #17 gives an example and discusses this further.) Consider your responses carefully. Don't exaggerate or use superlatives. Stay accurate.

- **Don't repeat a false or inaccurate statement.** Although the temptation is to assert that the (false issue) is not correct, remember that every time you repeat the statement, you have reinforced it in the mind of someone half-listening. Respond to the issue or allegation but without repeating it. (See Case Study #23 for an example).

- **Don't speculate.** Even though the interviewer asks a *"what if....?"* question, bring the topic back to what exists, what has been done, and what is scheduled. One way to do this is to draw attention to it: *"Well, that's just speculation, and your viewers/listeners want the facts. So here is what we've done in the last 7 days … here's the new system we'll have in place by this time next month,* etc." Most reporters will back off and stay on the facts.

- **Humour**: Don't use humour or sarcasm when responding to questions about serious health or environmental risks. Especially don't appear to be amused (even if it's just a nervous smile) when asked a serious question about a sensitive concern. That camera shot can be taken and used out of context.

- **Don't refer to the amount of money** that has been spent on fixing a health, safety, or environmental problem, at least at the same time that you are discussing the risks. It sounds as if you are putting a price tag on people's health and safety.

- **Don't state that zero risk is an impossible objective** to achieve. It's true in most cases, but that doesn't help people who are anxious about health and safety. Instead, insist that the efforts, programs, and activities are focused on moving toward a zero-risk *goal*.

- **Never get angry, appear flustered, or lose self-control** in an interview. It will always show you in a poor light.

- **Don't attack an opponent's credibility or character**. This is the ad hominem position, and it invariably lowers your own credibility. Always argue with the issue, the evidence, and the facts, not the person.
- **Low and slow**. In responding to questions, don't rush the response, drop your voice one tone lower one degree slower. This can give your response a slightly more confident aura and also allow you time to better choose and order your words.
- **Don't "ramble on"** in a live radio or TV interview. Ask the interviewer or producer about the kind of responses they are looking for in the segment. It may be 5- or 10-second sound bites for a street interview, but they may want longer reactions or explanations in a studio or zoom interview.
- **Repeat the key points** of your primary communication objective (PCO) if you have an opportunity. Each time, it sinks in a little more deeply to listeners or viewers.
- **Three points**. It's a good idea, where possible, to have your PCO described as three distinct points. This assures that all three will be reported; no editor will snip the recording after just one or two. And it also allows a chance of repeating the main reason (the previous point). For example:

"You ask why I am nervous about Ontario re-opening right now? There are three reasons: First, we are about to enter winter, which means indoor activities, lots of people sharing the same volume of air. Second, every time we have started to relax precautions, this virus has ambushed us with a new variant. And third, the UK relaxed too soon and their numbers are surging! That's why I'm nervous about Ontario re-opening right now!"

- **Don't assume you have been understood** by the reporter. ASK and verify that they have understood, and clarify if necessary. Remember **you** are the technical expert.
- **Don't talk in theories and abstractions** when you can use examples, comparisons, or analogies.

"The spectrometer we'll be bringing in is about the size of a microwave oven...."

- **In a longer-form interview, use stories and anecdotes to illustrate a point**. This is not a scientific conference. People appreciate stories about problems being solved and examples of how a person was helped or a change was made. Personal accounts are even better.
- **Don't compare unrelated risks**. Even if the lifetime risk of cancer from water containing perchloroethylene *is* statistically the same as being hit by lightning, the comparison is being made between a man-made risk and a natural risk, and it will sound false. Always compare similar or equivalent risks.
- **Never ask to be trusted**. Recall who in your experience says *"Trust me."* You can think immediately of certain politicians, or perhaps someone selling used cars.[3] If you need to *ask* to be trusted, be prepared for a reaction that perhaps you are slightly suspect. Trust is earned not by speeches but by repeated actions, transparency, and accountability.

6.8 The "Town Hall" or Community Meeting

6.8.1 Make Sure All Groups and Stakeholders Are Invited

This is not the time to alienate subgroups. Ensure postal codes are correct and that information packages are delivered by hand if necessary and in sufficient time in advance of the meeting. Advanced notice in community newspapers should be posted at least a month before the meeting, with information mailed or emailed to participants 2–3 weeks ahead of the meeting.

6.8.2 Organizing and Planning the Meeting

Try to avoid the assembly hall or theater format in which "experts" on the stage, in suits, ramble on at great length explaining to the

---

[3] Author T.S. was making a presentation in the UK, and at this point a woman in the front row added "*and men!*"

audience "what they need to know" by means of interminable PowerPoint® lists, diagrams, and charts. This can rapidly become alienating and counterproductive if you have invited people with the express purpose of hearing about their concerns. A better arrangement can sometimes be a "flat" meeting area (school gymnasium, conference room) beginning with a short introductory session. Participants then break out to visit tables and posters around the walls and corners staffed by knowledgeable people who can answer questions, address concerns, etc., for 1–2 hours, with a final "plenary" session. A horseshoe arrangement for seating sometimes works better than with a podium and audience on different levels.

Some years ago, EPA communicator Susan Santos had been brought in to try to calm a community that had not shown a drop in anxiety after several town hall meetings. She requested more time and began door-to-door visits to the community residents. There, in peoples' own living rooms, her team members were able to listen to residents' concerns and provide the information they needed. The process successfully addressed the community anxiety.

6.8.3   Length of the Presentation

Any formal presentations should be as short as possible, with *most of the time given to people asking questions*, and those questions being answered or at least commitments made to get back to the questioner within a set period of time.

6.8.4   The Presentation at the Meeting

- **Watch for damaging nonverbal communication**. Your verbal content may not be as important to your credibility as your nonverbal actions. Body positioning can be expressive without you realizing it. You've just finished telling attendees that the whole purpose of this meeting is to listen carefully to their questions, but as the first person starts to speak, you turn your back and fumble with papers on the desk behind you.

- **Hand gestures** are important, especially when they contradict your spoken words. The open hand, palm outward, can superficially appear to be conveying "... *Hold on, there's more ....*" But many will interpret the action as "... *Sit down, I'm speaking!*" This is not helpful in an environment where you are trying to reassure the audience and generate credibility.
- **Clothing**: Avoid the suit and tie. You'll be more believable in working clothing. Rolled-up shirtsleeves or surgical scrubs are also less confrontational than a business suit.
- **Use graphics in visual presentations**. Attention span will not survive point-form slides for long; employ *simple* designs and shapes, *uncluttered* charts and diagrams, and colors as much as possible.
- **Don't insist** *"This is what you need to know!"* Listen closely to what people are saying, and address their questions. That way you will find out what is concerning them.
- **Claiming** *"We need to hear your concerns!"* backfires if it turns out that no-one is taking the minutes, or recording the session, and especially if there's no follow-up.
- **Humor** can be helpful in most relaxed presentations, BUT be very careful. Don't use humor when your audience has concerns about serious health or environmental risks.
- **Listen.** You are at a meeting with anxious community members to listen to their concerns and questions. This is NOT the time to deliver a lecture! Rearrange the letters of the word *LISTEN*; you will discover the word *SILENT*. Take the hint. This is the time to let *them* talk, to let them tell you their worries and concerns, and to ask their questions. That's when you can say something informative and helpful.
- **Don't be overconfident or overassertive**. Include uncertainty, and explain if some of the details are incomplete. Include any personal uncertainties and concerns you may have. And above all, at some stage, ask yourself: *"Would I really want to live here with my family and drink the water on a daily basis?"* If you have difficulty with the answer, why are you trying to convince these people that they have nothing to worry about? (See also Case Study #15 below.)

- **"Layering"** is a communications technique that simultaneously presents the information in more than one format such that more complex details can be appreciated by those with appropriate understanding, while a layer or two of less-complex information finds a ready reception by those who perhaps lack scientific or technical understanding. The object is to avoid both "talking down" to some people and also avoid going over heads of others with what seems to them like jargon.

- **Empathy** is key. Your audience is not likely to be impressed by your degrees, published papers, or professional reputation. They *will* closely observe your presentations and discussions, watching for indicators of caring, empathy, ability to listen, ability to see issues from the perspective of other people. These characteristics were found by Covello (1995) and others to be the most important predictors of community trust and credibility.

- **Be sensitive to the perception of "individual" risk**. While agencies and epidemiologists prefer to measure, understand, and describe risk in terms of the threat to society and the population as a whole, the individual is focused firmly upon the risk to themselves and their family. This is a well-recognized dichotomy. An expressed incremental risk of one in a million (1×10^{-6}) appears not to exceed normally acceptable levels of risk, but to an individual in a city of 5 million inhabitants, the perception is that 5 deaths are expected and "casually written off" by the assessment and that someone in *their* family might be one of those five.

6.8.5   Following Up on Commitments

Minutes or records of meetings should be taken, and be *seen* to be taken, and *made available promptly* to all stakeholders through Internet channels, email, or hard copy for people not equipped with electronic technology. Questions left on the answering machine or online that remain unanswered a week later and certainly by the time the next community meeting takes place will not convince anyone of your sincerity or credibility. You've lost their trust. Follow-through is essential, and great effort must be made to respond to queries, questions, and requests for further explanation. Above all, don't hide behind a thinly veiled implication that the material or technology is *"complex and above the comprehension of the average citizen."* With creativity and effort (and experience), virtually anything can be explained in terms that are understandable.[4]

6.9   The Seven Questions to Prepare for Any Risk Communication

Good risk communication can and should be planned ahead of time, even, obviously, in the absence of the actual context or threat. This checklist, condensed from an article by Sly (2000), is a framework for risk communication response that will be valid and useful regardless of the nature of the issue.[5]

#1: Who should we tell?
- We tell all segments of the community the same info.
- But tell the people most affected first.[6]

#2: What should we tell them?
- Tell them what is known, more rather than less.
- Tell them what is not yet known.
- Tell them the extent to which the information may be uncertain, unreliable, or incomplete.

#3: What if the information is incomplete or doubtful?
- Tell them what you are unsure of.
- Promise to share all further information as soon as it is received.

---

[4]This principle is a key message in Lundgren and McMakin's (2018) handbook: *Risk Communication.*

[5]This paper has been adopted by the government of New Zealand and the US Dept. Health and Social Services as a training document.

[6]Peter Sandman calls this the "double indignity" of risk communication: the people most at risk are often the last to find out due to the socioeconomic and employment circumstances in which they live.

#4: When should we tell them?
- Tell them as soon as the agency knows,
- And <u>before</u> the local TV station breaks the story at 6 pm.

#5: How should we tell them?
- Respect their concerns.
- Explain so as to be understood.
- Avoid technical jargon and very large and very small numbers.
- And *rehearse* the message for clarity.

#6: Who should tell them?
- Must be technically credible and responsible
- Must be experienced at media relations, public speaking, and listening.

#7: Have we formed a partnership with the community?
- Essential throughout for stabilizing the concerned community.
- Must be genuine; tokenism can have the opposite effect.

6.10   An In-Depth Look at Communication Delayed

Item #4 in the list (Fig. 6.6) is arguably the most frequently ignored factor, and the source of most large-scale public anger, frustration, and (inevitably) loss of trust in the agency. For examples, see the BSE/vCJD saga (Case Study #19), the Belgian contaminated animal feed episode (Case Study #20), and the Wisconsin waterborne cryptosporidium outbreak (Case Study #21).

The recurring errors can be illustrated by means of a Punnett Square (Fig. 6.6), a device borrowed from genetics. The vertical discrete variable is that the risk will eventually be found to be "real" or "not real," while the horizontal discrete variable is the decision to act NOW to announce the issue or threat to the public, or to wait until "verification/confirmation" arrives. The four cells are the outcomes for each combination of factors:

Curiously, most of the time, government or corporate decision-makers fear cell #3. They are desperate to avoid "*upsetting people*

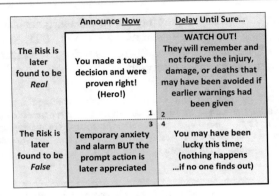

	Announce **Now**	**Delay** Until Sure...
The Risk is later found to be *Real*	You made a tough decision and were proven right! (Hero!) 1	**WATCH OUT!** They will remember and not forgive the injury, damage, or deaths that may have been avoided if earlier warnings had been given 2
The Risk is later found to be *False*	3 **Temporary anxiety and alarm BUT the prompt action is later appreciated**	4 You may have been lucky this time; (nothing happens ...if no one finds out)

Fig. 6.6 Analysis from delayed notification

unnecessarily should it turn out to be false!," thus choosing to *delay* and not to inform the public "*until they know for sure!*" (cell #2).

If, after delaying, the threat turns out to be true (cell #2), this quickly becomes the worst possible outcome, destroying trust, credibility, reputation, and careers. This will be remembered for a long time and could be grounds for litigation. If they had acted immediately, but the threat was false (cell #3), the anxiety soon subsides, and the action is usually not only *forgiven*, but in hindsight *much appreciated* for having been an appropriate response *had the threat been "real."*

The unfortunate tendency to delay announcement of impending risks or issuing timely advice has a long history and countless examples illustrating the folly of the practice. In modern times, the "precautionary principle" is often cited as the preferred rationale, and in Canada, two large scale inquiries have underscored its importance. In the final report of the Commission of Inquiry into failure of the Red Cross blood donation system in Canada to protect the blood supply from HIV (1997), Mr. Justice Krever said:

> "*Where there is reasonable evidence of an impending threat to public health, it is inappropriate to require proof of causation beyond a reasonable doubt before taking steps to avert the threat.*"

Similarly, in the final report into the way Ontario mishandled the severe acute respiratory syndrome (SARS) in 2003, Mr. Justice Campbell was critical of the way the Ontario government waited until definitive proof was assured:

> *"If the Commission has one single take-home message it is the precautionary principle that safety comes first, that reasonable efforts to reduce risk need not await scientific proof. Ontario needs to enshrine this principle and to enforce it throughout our entire health system."* SARS Commission (2006)

6.11   Case Studies in Risk Communication for Illustration and Discussion

Each of the following illustrates approaches and solutions to communication problems. They are all drawn from or based upon actual case histories, some exemplifying an effective response, while others present an opportunity for introspection, analysis, and redesign.

Case #	List of case studies
14	Interpreting a risk of 10^{-6}
15	Admitting uncertainty can actually increase trust
16	From "No way!" to "Maybe"
17	"Off the record" (no such thing!)
18	Interview only when you're prepared
19	BSE and vCJD in UK:
20	Dioxin in Belgium, 1999
21	400,000 ill in Milwaukee
22	Johnson & Johnson
23	Don't repeat false statements
24	Apologize promptly and apologize sincerely.

6.11.1 Case Study #14: Helping to Interpret 10^{-6}

Case Study #14: Helping to Interpret 10^{-6}

The community was angry and frustrated. They had learned that their water supply contained trace amounts of chemical that gave them a risk calculated at close to 1×10^{-6} or one in a million additional deaths from cancer over a lifetime.

At the town hall meeting, EPA Communicator Martha Bean suspected that these people were not necessarily familiar with scientific notation. This was her approach as she started to explain:

- "*Let's consider this group here tonight – it's about 200…*
- "*Forget about the water for a moment. For these 200 people, regardless of what water we drink, we can safely predict that around 25% of us – that's about 50 people here – will die of cancer…*
- "*Now let's suppose everyone drinks 2 quarts*

What can we learn from this?

Reduce the very large or very small numbers to a manageable size: between "thousands" and "thousandths" is familiar territory to most people.

Compare one risk with another (similar) risk. Consider contrasting the incremental risk against the background risk.

Very often (as here), the background risk is far greater. Without that comparison, people are left to focus intensely upon the small incremental risk.

(≈2 liters) of this same water every day for a lifetime; we can predict that there's only a 1 in 5,000 chance that just one more person might die of cancer over their lifetime...."

This approach doesn't of course guarantee that the citizens will immediately approve, but when the risk is explained in this way, it helps people to understand the risk.

This approach helps people see the problem in more manageable numbers. (The range thousands to thousandths seems more familiar than large exponential numbers.)

It also helped to contrast the risk from the water supply to the risk from all other sources over a lifetime. *A comparison like this gives a valuable point of reference, which was missing in a "stand alone" risk estimate.*

6.11.2 Case Study #15: Uncertainty Can Increase Trust

Case Study #15: Uncertainty Can Increase Trust

Consider a news bulletin in which the local medical officer of health announces:

> *"An hour ago it was reported that the town's water supply may have been contaminated and therefore is unsafe. Let me assure you that the water supply in this town has a flawless record for safety! It's clearly a rumor without any truth, probably a reporting error! No contamination has taken place!"*

Question: *What percentage of people listening to this announcement would believe the message unconditionally?* 30%? 40%? A good proportion would wonder about the validity of the message; some might suspect they were not hearing the whole story.

Now consider this alternative news bulletin:

What can we learn from this?

Overconfident communications and reassurances often backfire.

Be open and transparent about uncertainty – even tending to admit to slightly more uncertainty – with the assurance that you will follow through immediately as soon as the results are available!

Genuine uncertainty can be more credible in a transparent context.

But be sure to follow through as promised!

"An hour ago it was reported that the town's water supply may have been contaminated and therefore is unsafe. Our field staff are taking samples as you hear this, and will be taking another 100 samples to the lab this afternoon. The lab will also work around the clock on this. I will report to you immediately when we have the results, but in the meantime, I want you to boil all water for drinking unless you are making tea, coffee, or soup, just in case. We think it's a reporting error, but to be on the safe side, please boil your water briefly before you drink it"

And again, the question: *What percentage of people listening to this announcement would believe the message unconditionally?* You would probably expect 100% of everyone listening would believe the announcement. After all, there's nothing to generate doubt or suspicion.

Notice that in our thought-experiment, the very assertive, very confident announcement was unlikely to be believed by everyone, whereas the announcement that introduced uncertainty would probably be believed by most, if not all people.

6.11.3 Case Study #16: Getting to "Maybe"

Case Study #16: Getting to *"Maybe"*

A community has just been informed that a large company is planning to build a large composting/recycling plant at the edge of the town.

What can we learn from this?

The more collaborative approach of "B" empowered the community by introducing a form of partnership and ultimately a voluntary, informed decision. It has brought the community to **"maybe,"** *which is a long way from the lengthy and bitter resistance expected in "A." The costs to the company are negligible in money and time. But if they do establish the plant, they are months or years ahead of the "A alternative.*

In terms of the outrage factors, they have avoided or at least reduced several: coercion, local lack of knowledge, local lack of control, lack of trust, and lack of responsiveness.

Scenario A: The corporation announces that they have carried out an assessment and this site is their choice. The community vigorously resists, on the grounds that they were neither consulted nor even told that such a development was being considered in their quiet and pleasant community. Local residents prepare to dig in and oppose. Resistance may rage on for years.

Scenario B: The recycling company announces that this site is their preferred location. The company offers the community a one-time technical services grant of $60,000, which will allow the community to hire their own expert on facility siting. "Once you have the expert," says the company, "we will sit down with the community and the consultant, and we'll talk about the process, the type of environmental monitoring, who does the monitoring, which laboratory will analyze the samples, and the maximum acceptable limits for air and water sampling. We'll also discuss liability, insurance, compensation, and the point at which the plant will cease operation if limits are exceeded."

The company then tells the community that if they can agree on these matters, they will build their plant as planned. If not, they won't, and they are prepared to make this promise in writing.

Nothing in scenario B guarantees that the community will approve the installation, but unlike scenario A, the community is much more likely to proceed with acquiring a consultant and entering into discussions. What do they have to lose? And thinking ahead, they also realize that once the monitoring and precautions are all in place, with some community oversight and even veto power, there may also be benefits; a new water supply, perhaps, or some better roads, a stronger local tax base, and of course employment – jobs will be available.

[This example is based on one of the many scenarios offered by Peter Sandman (2012a, b), whose websites, seminars, and publications are valuable sources of experience, wisdom, and examples about communicating risk. https://www.psandman.com/col/4kind-1.htm]

6.11.4   Case Study #17: Off the Record

Case Study #17: "Off the Record"

The interview with a local TV broadcaster had been about food safety issues, the local food hygiene inspection program, and restaurant standards in general.

The interview (with author T.S.) was declared over, and while the cameraman busily coiled some wire, the reporter thanked the interviewee for the background information about inspection and enforcement of food hygiene regulations.

What can we learn from this?

If there's a reporter in the room or on the other end of a phone call, or a camera or a microphone is present, you cannot consider anything to be "off the record."

The reporter put aside his notes, and leaned in a little closer, saying: *"You've had a lot of field experience in this work; I'll bet you've seen some things, and have lots of stories about really bad places!"*

He paused, presumably anticipating some juicy morsels about disgusting scenes in restaurants, naming names, and sharing behind-the-scenes gossip about specific local eateries.

The unattended camera, however, was still mounted on its tripod with the red recording light "on." Perhaps an oversight? Perhaps not. But there's no doubt that had the author been careless enough, or unprofessional enough to reveal anything at all about local businesses, THAT is what would have found its way onto the 6-o'clock newscast.

6.11.5 Case Study #18: Interview with Information

Case Study #18: Interview with Information

The common black mold *Stachybotrys chartarum* was causing some public concern due to a recent story from Cincinnati where the mold had been declared toxic and perhaps implicated in the deaths of several infants. This was not subsequently confirmed, but at the time the public was alarmed, and the health department had agreed to carry out inspections of several high-rise apartment buildings to assess how much *Stachybotrys* mold had grown in and on walls of poorly ventilated apartment units.

What can we learn from this?

Do not agree to the interview unless you have all the available, current, details. Explain to the reporter that you'll do the interview but you want to be sure you have the up-to-date facts. Ask them to return at 9 am. They will not be happy but they will be there.

Go through the information with your team to make sure you are current and as complete as possible. And rehearse responses to potential questions, for the best, clearest, way of getting the message across.

Remember that although you are the "content" specialist, the reporter is trained to find the stories.

The local TV news team called to arrange an interview with the associate medical officer of health (AMOH), who agreed to the interview. Unfortunately, the AMOH was not familiar with the inspection schedule, where the inspections had been completed to date, and what had been found.

The interview was an embarrassment due to the incomplete information and incorrect facts. The reporter had already interviewed a resident; when the piece showed on prime time, footage of the resident's observations was alternated with the responses by the AMOH. The two accounts were completely contradictory.

As viewers watched, the focus of the story quickly changed away from the "mold problem" to the issue of the associate medical officer of health who appeared to be unaware of what was going on in the health department.

6.11.6 Case Study #19: The BSE Risk Communication Failure

Case Study #19: The BSE Risk Communication Failure

It was September 1985 and Veterinarian JM Watkin-Jones read the lab result from a sample of bovine brain tissue he had submitted to the laboratory from a farm in Winchester (UK): *"Spongiform encephalopathy."* It was the first identification of Bovine spongiform encephalopathy (BSE), a new illness that was to affect millions of cattle in the UK. The animals were observed with unusual symptoms, unsteady gait, and inability to stand after falling. As the beef industry began to decline, the British Ministry of Agriculture, Fisheries, and Food (MAFF) rushed to assure the population, announcing repeatedly:

"…this is an animal disease, you are not at risk, trust us."

What can we learn from this?

For a decade, the public had been assured of the safety of British beef without adequate scientific evidence to do so.

When the transmission to humans was finally confirmed, the public expressed strong resentment that public safety had been compromised for commercial interests and trade interests.

Trust was once again the major casualty. MAFF was replaced by a different ministry.

The precautionary principle had been ignored completely at the expense of public health.

For the next 10 years, MAFF insisted that British beef was safe, despite growing evidence that it was not limited to cattle. The source of transmission was becoming clear: animal feed, including milk replacer, began as "meat and bone meal" (MBM), powdered bovine, and other mammalian tissues from culled and diseased animals. In addition to cattle-fed MBM, cats, both domestic and large (e.g., white tigers), became infected and died after being fed meat that was not fit for human consumption. Scientists admitted being worried in case humans could be infected, especially as the incubation period might be a decade or more. The agent was found to be an infectious protein, a "rogue" *prion*, placing the disease as one of a family of transmissible spongiform encephalopathies (TSEs) including Kuru and CJD in humans, and several others, including chronic wasting disease (CWD) and scrapie, in animals. The government still insisted that humans were safe, right up to March 21, 1996, when the MAFF admitted that the first ten cases had been found of what appeared to be BSE having entered the human population. The new human disease was named *variant Creutzfeldt-Jacob Disease* (vCJD).

During those 10 years, it is estimated that between 500,000 and a million infected cattle had been processed and eaten in cheaper beef products such as pies, burgers, and sausages. The public henceforth distrusted all information coming from MAFF, and the function of that ministry was later taken up by a new food safety agency (Webster et al., 2010).

6.11.7   Case Study #20: Dioxin in Animal Feed

Case Study #20: Dioxin in Animal Feed (Belgium)
The crisis first came to public notice on May 28, 1999. Concentrations of dioxin were found in chickens and eggs sold in Belgium. These products were banned from sale. Early the next month, the ban was extended to include a much wider range of foods including prepared foods.

It was subsequently established that the contamination began *5 months earlier* in January 1999, when animal feed companies incorporated oils and fats, probably contami-

What can we learn from this?

The delay by the government in acknowledging the incident and responding responsibly in the interests of the public's health exceeded 2 months.

This was perceived as a cover-up, and resulted in great loss of trust and credibility.

nated by unknown materials, which they had received from their supplier.

In February, ill chickens had been observed on a number of poultry farms, and on March 18, 1999, laboratory tests on some of the products were made. On 26th of April, the results of the tests were released to the Ministry of Agriculture. The tests had shown the contaminant was dioxin.

The contamination had indeed spread much further than the animals that had initially been fed, and now extended to beef, poultry, milk, eggs, butter, cheese, and foods manufactured from them. Both the suppliers of the raw materials and the animal feed companies were found to have delayed notifying the authorities, and the Belgian authorities also responded with long delays before the first measures were taken and the public was informed.

Source: Council of Europe: Food Safety Committee Hearings (1999)

6.11.8 Case Study #21: 400,000 Ill in Milwaukee

Case Study #21: 400,000 Ill, 69 Dead in Milwaukee
Early in the spring of 1993, an outbreak of diarrheal ill-ness was reported among the citizens of Milwaukee, Wisconsin. By the time it was ended, approximately 400,000 cases and at least 69 deaths had been recorded. For

What can we learn from this?

This episode, nearly 30 years ago, demonstrates that if any factor (here, the water) is suspected as a possible cause of serious harm, the appropriate measures to prevent further harm should be implemented without waiting for final "proof."

This is the *"precautionary principle."* It is also reflected in Communication Case Studies 19, and 20, and is the basis for item #4 in the Seven Fundamental Questions in risk communication (Sect 6.9).

2 weeks, newspapers ran stories on what might have been causing the mystery disease. Water is always suspect when a sudden "explosive" outbreak occurs, but investigators were assured that the two water purification plants were treating their water as designed and that the chlorine levels were correct. The focus turned elsewhere: a new virus, food, air, pets?

The assurance that water treatment was up to standard appears to have obscured the investigation, until the Milwaukee City Laboratory director, Steve Gradus, called William R. Mac Kenzie, the CDC epidemiology intelligence officer, to report he had found *Cryptosporidium* parasites in several fecal samples. (*Cryptosporidium* was considered a newly described organism and was poorly understood at the time.) Mac Kenzie considered this to be an important finding. Cryptosporidiosis is spread through water, it causes watery diarrhea, it cannot be detected with standard water quality tests, *and it is resistant to the levels of chlorine* normally used for removing bacteria.

Mac Kenzie met with the mayor and chief epidemiologist and told them the *Cryptosporidium* in the water supply could be the reason for the outbreak and suggested an immediate boil-water order to all residents. But there was hesitation in issuing a boil-water order because of the city-wide scale of the order and the risk of scalding, and because it was based on preliminary information that had not been confirmed.

At the meeting, the mayor asked a State Department of Health official if he would drink the water knowing that information that they had learned.

"*No*," said the official.

The mayor made a decision. "*If you wouldn't drink the water, I don't want the people of Milwaukee to drink it either.*" The boil-water order was given and the epidemic quickly diminished.

6.11.9 Case Study #22: Johnson & Johnson: A Crisis Handled Effectively

Case Study #22: Johnson & Johnson: Demonstrating Responsiveness

Tylenol, Johnson & Johnson's trusted brand of acetaminophen, accounted for 37% of the market share by early 1982. For reasons unknown, during the fall of the year, a person or persons filled capsules with calcium cyanide, put them into containers of Tylenol Extra Strength, and placed them on the shelves of several food stores and pharmacies in the South Chicago area. Seven people died as a result. J&J was faced with the task of explaining why its trusted product was suddenly harming people.

What can we learn from this?

Reveal the issue immediately; don't wait for 100% confirmation.

Advise the affected population about the hazard, the precautions they must take and the steps you are taking, without delay, and update promptly.

Respond with transparency and without hesitation, even if it is painful!

James Burke, Chairman of J&J, formed a seven-member strategy team with the primary task of "protecting the people" and the secondary task of "saving this product."

The company immediately ordered the withdrawal and destruction of all Tylenol product in capsule form from store shelves in the South Chicago district, but when two more contaminated bottles were found, the company instigated a national recall of Tylenol. A public news release from the company informed consumers not to use *any* Tylenol product they might have at home, until the extent of problem had been identified.

Within hours, J&J set up a 1–800 line for consumers to call, a toll-free line for news organizations to call and receive updated statements, and a live television feed via satellite allowing all press conferences to go national.

Soon after, J&J presented their new triple-safety-seal packaging, the first product in the industry to use the new tamper-resistant packaging.

> **Question:** *Had the recall and destruction of 17 million dollars' worth of product across the USA been ordered with the objective of reducing the **actual** risk? (The tampered product had only been found in a local South Chicago neighbourhood). Or was it to establish nationwide responsiveness and genuine concern for the well-being and protection of their customers?*

The company's handling of the crisis is seen as an example of how major business should react, respond, and communicate with the public during a crisis. They made public safety the top priority, appropriate for a company whose product line and reputation had been established on maintaining health.

Within a year, the Tylenol brand name had regained its reputation as one of the most trusted over-the-counter preparations in the USA. Before the crisis, Johnson & Johnson had little experience with media relations, in crisis mode, but the company quickly recognized the value of rapid, open, responsive communications with the public.

6.11.10   Case Study #23

Case Study #23: Don't Repeat a False Statement

Faced with a false accusation or provocative statement, the worst response includes repeating the false statement. Here's an example:

Reporter's Q: *"We're hearing a story this morning that a food handler in the hospital is a carrier of typhoid fever bacteria. How do you respond?"*

Response A: *"A food handler carrying typhoid? There's no food handler carrying typhoid in this hospital, and I'll tell you why – because all hospital food handlers are tested when they begin, and yearly! That's why it's completely false to claim a food handler is carrying typhoid!"*

Someone listening with half an ear while they drive to work has just heard the original claim, followed by YOU repeating three more times the words *"...food handler carrying typhoid ... ... food handler carrying typhoid ... ... food handler is carrying typhoid...."*

Is it any wonder that they arrive at their destination with the strong sense that someone at the hospital might be spreading something nasty!

What can we learn from this?

Pause before responding to give yourself a moment to plan.

In the same way we try to avoid a negative word, always try to avoid repeating an untrue or provocative statement.

Practice this, and take advantage of workshops and courses to improve your risk communication skills.

Practice a response that avoids repeating the claim:

Response B: *"All hospital food handlers are tested when they begin employment and each year after that. We are looking into that story right now. The hospital food service is operating with extra precautions, and we have arranged to have all food handlers retested in the next 3 days. I'll update you as soon as I have the results."*

And to illustrate that such situations do occur, here's an example of an awkward statement made in March 2003, when the SARS-1 virus (at the time referred to as "atypical pneumonia") was beginning to spread. It's from the Secretary for Health for Hong Kong at the time, Dr. E.K. Yeoh:

"Hong Kong is absolutely safe and no different from any other big city in the world…. Hong Kong does not have an outbreak, okay? We have not said that we have an outbreak. Don't let the rest of the world think that there is an atypical pneumonia outbreak in Hong Kong."

Dr. Yeoh was highly respected as a physician and public health expert, but a momentary response, a careless communication, can easily give a wrong impression.

6.11.11   Case Study #24

What can we learn from this?

° **Don't hesitate** to acknowledge the error. The time frame is hours not days or weeks.

° **Don't shift the blame.** Accept the responsibility.

° **Apologize sincerely** and keep apologizing until the groundswell of response begins to reflect:
"We've heard your apology, now let's get to work".
...But that must *never* be your call.

Case Study #24: Apologize Promptly; Apologize Sincerely: Three Studies

Agencies, companies, and individuals are typically reluctant to apologize for mistakes or errors. Legal advice at an earlier time had warned against saying "sorry" or admitting guilt, ostensibly to prevent an avalanche of claims, but that has changed. The company or agency that fails to *admit fault sincerely*, and *apologize immediately* is quickly perceived in a negative light and faces a rough road to regain their former position and reputation.

In 1989, the *Exxon Valdez* oil tanker ran aground off Alaska and spilled more than 11 million gallons of crude oil into the sea and onto surrounding shoreline, devastating the ecosystem. Up to that time, it was the worst oil-spill disaster in the USA. The company remained mostly silent, but on the tenth day instead of issuing a personal apology, the CEO ran advertisements in newspapers, expressing regret, but, in the view of the public, the media, and even oil-industry observers, did not clearly accept responsibility. At the same time, the company threatened to raise oil prices to pay for the damage and cleanup.

Twenty-one years later, the Macondo oil well, 5000 feet below the surface, 250 miles southeast of Houston, Texas, ruptured and spewed an estimated 184 million gallons of oil into the Gulf of Mexico. BP (formerly known as the British Petroleum Company) responded, apologized, and ultimately paid out a reported $60 billion in costs and reimbursements. They avoided some of the earlier errors of Exxon's response, but the sincerity was lacking. On May 30, 2010, the frustrated chief executive stated, *"We're sorry. We're sorry for the massive disruption it's caused to their lives. And there's no one who wants this thing over more than I do. I'd like my life back."*[7]

In contrast, on August 23, 2008, Canada's Maple Leaf Foods was found to have caused illness and death through Listeriosis. A sombre CEO, Michael McCain, appeared on television the same evening: *"When Listeria was discovered in the product, we launched immediate recalls to get it off the shelf, then we shut the plant down. Tragically, our products have been linked to illness and loss of life. To Canadians who are ill and to families who have lost loved ones, I offer my deepest sympathies. Words cannot begin to express our sadness for your pain. Maple Leaf Foods is 23,000 people who live in a culture of food safety. We have an unwavering commitment to keeping your food safe with standards well beyond regulatory requirements. But this week our best efforts failed and we are deeply sorry. This is the toughest situation we have faced in a hundred years as a company. We know this has shaken your confidence in us. I commit to you that our actions are guided by putting your interests first."*[8]

---

[7]From BP CEO Tony Hayward: "I'd Like My Life Back," YOUTUBE (May 31, 2010). A clip of statement made on The Today Show (NBC television broadcast May 30, 2010). http://www.youtube.com/watch?v=MTdKa9eWN Fw&feature=related

[8]The company used a television announcement to apologize and to explain its response. The 66 second video was broadcast widely and posted on YouTube on August 23, 2008.

References

Council of Europe: Food Safety Committee Hearings. Dioxin crisis and food safety. Report Doc. #8453. Committee on Agriculture and Rural Development. 22 June 1999. https://assembly.coe.int/nw/xml/XRef/X2H-Xref-ViewHTML.asp?FileID=8023&lang=EN. Accessed July 2022.

Covello VT. Risk perception and communication. Can J Public Health. 1995;86(2):28–31.

Covello V, Sandman PM. Risk communication: evolution and revolution. In: Wolbarst A, editor. Solutions to an environment in peril. Baltimore: Johns Hopkins University Press; 2001. p. 164–78.

Diamond J. The world until yesterday: what we can learn from primitive societies. Viking Adult; 2012.

Fishhoff B. Risk perception and communication unplugged: twenty-years of process. Risk Anal. 1995;15(1):137–45.

Frewer L, Howard C, Hedderly D, Shepherd R. What determines trust in information about food related risks? Underlying psychological constructs. Risk Anal. 1996;16(4):473–85.

Hance BJ, Chess C, Sandman P. Improving dialogue with communities: a risk communication manual for government. Trenton: New Jersey Department of Environmental Protection. Division of Science and Research; 1988.

Hance BJ, Chess C, Sandman PM. Setting a context for explaining risk. Risk Anal. 1989;9(1):113–7.

Kasperson RE, Renn O, Slovic P, et al. The social amplification of risk: a theoretical framework. Risk Anal. 1988;4(8):177–8.

Krever H. Final report. Commission of Inquiry on the Blood System in Canada, vol. 295. Ontario Government Publications; 1997.

Löfstedt R. Communicating food risks in an era of growing public distrust: three case studies. Risk Anal. 2013;33(2):192–202.

Lundgren RE, McMakin AH. Risk communication, a handbook for communicating environmental, safety, and health risks. 6th ed. Hoboken: Wiley; 2018.

NRC (National Research Council). Risk assessment in the federal government: managing the process. Washington, DC: National Academy Press; 1983.

NRC (National Research Council). Science and decisions: advancing risk assessment. Washington, DC: National Academies Press; 2009.

Pew Research Center. Distrust, discontent, anger and partisan rancor: the people and their government. Survey reports. 2020. Available at: http://www.people-press.org/report/606/trust-in-government. Accessed 4 Jan 2011.

Powell D, Leiss W. Mad cows & mother's milk. Montreal: McGill-Queen's University Press; 1997. p. 26–40.

Sandman P. Risk = hazard + outrage: a formula for effective risk communication. Parts 1 to 5. Instructional video recordings commissioned by the American Industrial Hygiene Association. Revised 2012a.

Sandman P. Responding to community outrage: strategies for effective risk communication. First published 1993 by the American Industrial Hygiene Association. Revised 2012b. ISBN 0-932627-51-X. Full text online at: http://psandman.com/media/RespondingtoCommunityOutrage.pdf.

SARS Commission. Final report. Executive summary: Ontario Government Archives. Vol. 1. Dec 2006. p. 13–14. Available at: http://www.archives.gov.on.ca/en/e_records/sars/report/v2-pdf/Vol2Intro.pdf. Accessed Nov 2022.

Shannon CE, Weaver W. The mathematical theory of communication, vol. 1. Urbana: University of Illinois Press; 1948.

Slovic P. Perception of risk. Science. 1987;236(4799):280–5.

Sly T. The perception of risk in the sensitized community. Doctoral dissertation, Teesside University, UK. 1997.

Sly T. The perception and communication of risk: a guide for the local health agency. Can J Public Health. 2000;91(2):153–6.

Starr C. Social benefit versus technological risk. Science. 1969;165:1232–8.

Webster A, Conor MWD, Sato H. BSE in the United Kingdom. In: Sato H, editor. Management of health risks from environment and food. Dordrecht: Springer; 2010. p. 221–65. https://doi.org/10.1007/978-90-481-3028-3.2010.

Correction to: Assessment and Communication of Risk

Eric Liberda and Timothy Sly

Correction to:
E. Liberda, T. Sly, *Assessment and Communication of Risk*, https://doi.org/10.1007/978-3-031-28905-7

The initial incorrect values of certain elements in Chapters 1 and 2 have been corrected retrospectively to prevent confusion among readers. The correct information is given below:

Chapter #	Page	Line	EXISTING	CORRECTED
1	20	[B]L1	"The annual risk …"	"The risk …"
1	20	[G]L8	"There were 47 falls …"	"There were **44 falls** …"
1	22	C	"1.8 deaths /3600 workers per year = 1.0×10^{-4}"	1.8 deaths /3600 workers per year = $\mathbf{5.0 \times 10^{-4}}$
1	22	F	H/pilot risk: 0.0050 Fixed-w/pilot risk 0.0025; H/p twice the risk of death	**(a) 8×10^{-6}** **(b) 3.2×10^{-5}** **(c) 4×10^{-5}**

The updated versions of the chapters can be found at
https://doi.org/10.1007/978-3-031-28905-7_1
https://doi.org/10.1007/978-3-031-28905-7_2

Chapter #	Page	Line	EXISTING	CORRECTED
1	22	G	(a) 8×10^{-6} (b) 3.2×10^{-5} (c) 4×10^{-5}	**OR:0.71:** **(Inverting:1.41)** Falls requiring orthopedic treatment were 41% more likely to have used new safety equipment. The numbers are very small, but the suggestion is that the new safety equipment does not provide better protection against more severe injury.
1	22	H	OR:0.71: Only 71% of falls using existing safety equipment required treatment (1.41): Falls requiring orthopedic treatment were 41% more likely to have involved the new safety equipment. The numbers are very small, but the suggestion is that the new safety equipment does not provide better protection against more severe injury.	**H/pilot risk: 0.0050 Fixed-w/pilot risk 0.0025; H/p twice the risk of death**
2	37	L17	Visually, $12/272 = $ **0.12868**	Visually, $12/272 = $ **0.04412**
2	38	L5	$= 0.12868$	$= 0.04412$
2	45	(d)L26	Solution: $1- \{P(V_1^F$ and $V_2^F)^C\}$**15,600** $= 0.075$ or 7.5%.	Solution: $1- \{P(V_1^F$ and $V_2^F)^C\}$**15,600** $= 0.075$ or 7.5%.

Chapter #	Page	Line	EXISTING	CORRECTED
2	48	(f)	$P(C|T) = [0.9500]$	$P(C|T) = \mathbf{[0.99936]}$
2	67	Q(1)	[a] 11/16 = 0.6875, [b] 6/16= 0.375 …	[a] 11/16 = 0.6875, [b] **5/16= 0.3125** …
2	67	Q(9)	[a] 0.79515, [b] 0.00035, [c] 0.19005, [d] 0.20485	[a] 0.79515, [b] 0.00035, **[c] 0.01445, [d] 0.99965**
2	67	Q(18)	[a] 0.99, [b] 8×10^{-3}, [c] 7.912×10^{-3}, [d] 8.0×10^{-3}, [e] 8.0×10^{-8}	[a] 0.99, [b] 8×10^{-3}, [c] 7.912×10^{-3}, **[d] 7.99992×10^{-3} (round off to 8.0×10^{-3})**, [e] 8.0×10^{-8} **[f] 1.0×10^{-5} [g] 0.99999 [h] 0.002**
2	67	Q(20)	[a] 2.4×10^{-3}, [b] 11.84%, [c] with 3 gen risk: 0.25%, [d] liability drops from \$650 K to \$20 K	[a] 2.4×10^{-3}, [b] 11.84%, [c] with 3 gen risk: 0.25%, [d] liability drops from **\$947 K** to \$20 K

Glossary, Abbreviations, and Acronyms

ABS Proportion of substance absorbed during exposure

Absorbed dose Dose absorbed by the receptor before off-gassing and short-term excretion

Addition of probabilities (general rule) $P(A) + P(B) = P(A) + P(B) - P(A \text{ and } B)$

Addition of probabilities (special rule) $P(A) + P(B) = P(A) + P(B)$ where A and B are mutually exclusive

ADI Acceptable daily intake

Administered dose The initial exposure of the victim to the substance

ALARA (Principle) As low as reasonably attainable

AT Averaging time (denominator for intake calculation), in day units

ATSDR [US] Agency for Toxic Substances and Disease Registration

Background risk Estimates of the normal incidence among the whole population

Bioconcentration Factor extrapolating from medium (e.g., water) to consumable (e.g., fish)

[C] Concentration, commonly reported in units of mg/L (in water), mg/kg (in food), mg/m^3 (in air)

Carcinogen Substance or process capable of inducing or promoting any form of cancer

CDC US Centers for Disease Control (and Prevention)

© The Editor(s) (if applicable) and The Author(s), under exclusive license to Springer Nature Switzerland AG 2023
E. Liberda, T. Sly, *Assessment and Communication of Risk*,
https://doi.org/10.1007/978-3-031-28905-7

CDI Chronic daily intake or (I), with units mg/kg·d

Conditional probability The probability of A, given that B has (or will have) taken place: $P(A|B)$

CPF Carcinogen potency factor (see SF)

CR Contact rate, commonly reported in units L/d (water), kg/d (food), m^3/d (air)

De minimis "Level of risk that can be ignored." From common law maxim: *de minimis non curat lex* (the law does not concern itself with trifles). This has changed from time to time but currently, for most civilian purposes is 1×10^{-6}

Deterministic Calculation of risk using best-available single point-values (see also stochastic)

Dose response Step 3 in NRC model: quantifying exposure for the receptor (human victim)

ED Exposure duration, in units (y)

EF Exposure frequency, in units (d/y)

EPA Environmental Protection Agency (US)

Exposure assessment Step 2 in NRC model: determining pathways for exposure

Exposure point The place where the receptor (exposed individual) is exposed

FMEA Failure mode effects analysis

FTA Fault tree analysis

Hazard identification Step 1 in NRC model, determining priorities for subsequent assessment

HAZOP Hazard and operability analysis

HI Hazard index, a probability estimate of mortality from non-carcinogenic causes (no units)

Human health risk assessment A framework encompassing the technical and scientific information that relates to the risks and threats to human health and safety, including its organization and characterization. The typical endpoint is a statement addressing probability, that given exposures will harm human health (NRC 1983)

I Intake (see CDI)

Incremental risk The risk that can be attributed just to the exposure being studied

IRIS Integrated risk information system (US EPA)

ISO 22000 (International Standards Organization) Document # 22000 Food safety

ISO 31000 (International Standards Organization) Document # 31000: Risk Management

LADD Lifetime average daily dose

LC50 Concentration at which 50% of test animals die

LCDC [Canada] Laboratory Centres for Disease Control

LD50 Dosage at which 50% of test animals die

LOAEL Lowest observed adverse effects level

LOEL Lowest observed effects level

Monte Carlo Method of stochastic (probability) calculation of risk

MORT Management Oversight and Risk Tree analysis

Multiplication of probabilities (general rule) $P(A$ and $B) = P(A) \times P(B|A)$

Multiplication of probabilities (special rule) $P(A$ and $B) = P(A) \times P(B)$ where $P(B) = P(B|A)$

Mutagen Substance capable of inducing mutation to cellular DNA/RNA

NAS [US] National Association of Sciences

NOAEL No observed adverse effects level

NOEL No observed effects level

NRC [US] National Research Council

O.R. Odds ratio

PHAC [Can] Public Health Agency of Canada

PRA Probabilistic risk assessment applied to events, incidents, and catastrophes

Probability Must be expressible between 0 and 1. The likelihood of an event happening.

Q1* Alternate name for SF.

QRA Quantitative (chronic) risk assessment of outcome following exposure to contaminants.

RCA Root cause analysis

Retained dose The dose retained after excretion or exhalation

RfC Reference concentration

RfD Reference dose: usually 1/1000 of NOAEL

Risk Defined technically as $R =$ Magnitude $\times$ Probability

Risk characterization Step 4 in NRC model: calculating and explaining probability of outcome

Risk communication An interactive exchange of risk-related information among all stakeholders (individuals, groups, and institutions), including reactions to risk messages and about the structures and provisions for managing risk.

Risk management Risk management includes all the activities involved in prioritizing the risk, planning for its remediation, reduction, etc., and any other technical and nontechnical factors (e.g., finances, resources, policies, priorities, and many others), to reach a decision about the need for, and the extent of appropriate risk reduction measures, whether they are needed now, and the position they should occupy in the priority list of pending work. Solutions or other risk reduction measures also should target ways of achieving that reduction (NRC 1983). See also ISO 31000.

RME Responsible Management Entity (US EPA)

ROD Record of decision (US EPA)

RR Relative risk. Also called risk ratio.

RR Retained rate. Proportion of substance retained after initial off-gassing and other excretion

S Sample space, the list of all possible unique outcomes of a random trial or experiment.

Scenario Collection of personal/environmental factors specific to the assessment

SF (Also CPF, Q1*): Slope factor, with units [1/(mg/kg·d)] or [(mg/kg·d)$^{-1}$]

Stochastic Risk calculation from distribution of values, not from single point value.

Teratogen Substance capable of inducing gametocyte mutation and deformed fetus

Total risk Background risk + incremental risk

Toxicity Score Means of assessing/prioritizing on-site substances for detailed analysis later

Index

© The Editor(s) (if applicable) and The Author(s), under exclusive 247
license to Springer Nature Switzerland AG 2023
E. Liberda, T. Sly, *Assessment and Communication of Risk*,
https://doi.org/10.1007/978-3-031-28905-7

Printed in the United States
by Baker & Taylor Publisher Services